AMERICA'S CLIMATE PROBLEM: THE WAY FORWARD

AMERICA'S CLIMATE PROBLEM:
THE WAY FORWARD

Robert Repetto

publishing for a sustainable future

London • Washington, DC

First published in 2011 by Earthscan

Copyright © Robert Repetto 2011

The moral right of the author has been asserted.

Earthscan Ltd, Dunstan House, 14a St Cross Street, London EC1N 8XA, UK
Earthscan LLC, 1616 P Street, NW, Washington, DC 20036, USA

Earthscan publishes in association with the International Institute for Environment and Development

For more information on Earthscan publications, see www.earthscan.co.uk or write to earthinfo@earthscan.co.uk

ISBN 978-1-84971-214-9

Typeset by JS Typesetting Ltd, Porthcawl, Mid Glamorgan
Cover design by Rogue Four Design, www.roguefour.com

A catalogue record for this book is available from the British Library

Library of Congress Cataloging-in-Publication Data

Repetto, Robert C.
    America's climate problem : the way forward / Robert Repetto.
        p. cm.
    Includes bibliographical references and index.
    ISBN 978-1-84971-214-9 (hardback)
    1. Energy policy—Environmental aspects. 2. Renewable energy sources. 3. Petroleum industry and trade—Government policy. 4. Climatic changes—Government policy. I. Title.
    HD9502.A2R46 2010
    363.738'745610973—dc22
2010039472

At Earthscan we strive to minimize our environmental impacts and carbon footprint through reducing waste, recycling and offsetting our CO_2 emissions, including those created through publication of this book. For more details of our environmental policy, see www.earthscan.co.uk.

Printed and bound in the USA by Edwards Brothers.
The paper used is FSC certified.

Contents

Foreword

Timothy E. Wirth

The central challenges of this century are the subject of this book. Simply put, our planet and its life support systems are at a tipping point. The trends are not sustainable.

In 1987 the World Commission on Environment and Development, chaired by Dr Gro Harlem Brundtland, defined sustainable development as prosperity and progress that 'meets the needs of present generations without compromising the ability of future generations to meet their own needs'.

This timeless, universal statement of values perfectly captures the interdependence that characterizes the human experience today. People, problems and opportunities are connected as never before. Facing catastrophic climate change, as we do, it matters little whether greenhouse gases emanate from Chicago or Shanghai – we all get warm together.

The reality of this kind of interdependence is entirely new in human history. Never before have the fates and responsibilities of the world's people been as intertwined. Never before have we borne such heavy responsibilities toward our children and grandchildren.

Unhappily, we are not living up to those responsibilities. In fact, we are failing badly as stewards of the natural systems that make human life and planetary productivity possible.

Clear scientific evidence, accumulating daily, indicates that the economic, environmental and security costs of climate change will dwarf even the profound global effects of recent economic turmoil. But even as we better understand the science, our emissions are growing faster than predicted, the concentrations in the atmosphere are getting denser, and the time horizon in which we can effectively act is getting shorter, even as the risks grow larger.

It is dismaying, then, that my generation has been so utterly ineffective, so seemingly indifferent in the face of the established facts and the clear implications for the next generations. The world's future is being mortgaged away because we refuse to acknowledge how much our long term economic and national security is linked to the health of the planet's life support systems.

I have long believed that the fundamental obstacle to the pursuit of sustainable development – in the US and around the world – is the misguided belief that protecting the environment is somehow antithetical to economic interests. Too many of today's leaders will say: 'I'm for protecting the environment – as long as it doesn't cost jobs or hurt the economy.'

It is within this terribly mistaken analysis that we encounter the fundamental intellectual challenge to sustainable development. Over the long term, living off our ecological capital is a bankrupt economic strategy.

Stated in the jargon of the business world, the economy is a wholly owned subsidiary of the environment. Virtually all economic activity is dependent in some way on the environment and its underlying resource base – everything from food and fuel to water and fibre. These are the foundations of the vast majority of all economic activity and most jobs. When the environment is finally forced to file for bankruptcy because its resource base has been polluted, degraded and irretrievably compromised, then the economy goes down with it.

At the dawn of the 21st century, we find ourselves at a pivotal moment in the history of the world – one that requires the same kind of innovation, engagement and political will that our country has historically demonstrated – to transform the global energy economy.

This challenge of sustainability begins at home. To stabilize the atmosphere, we have to reduce our carbon output by at least 90 per cent by 2050. Each of us in the US is responsible for about 20 tons of carbon emissions per year, and most of it stays in the atmosphere for 100 years or longer. Instead of 20 tons, we will have to reduce our carbon emissions to about 2 tons per person. As a nation, we need to show what it means to transition to a clean energy economy and develop enlightened land use policies. We can set the standard for others to follow.

There are two relatively easy opportunities to accelerate this transition. First, a large scale effort to raise energy efficiency – better understood as energy productivity – can pay big economic dividends and deliver environmental and security benefits while reducing energy demand growth by half. Second, greater use of natural gas, which has only half the carbon content of coal and is much more abundant domestically than previously thought, is a low cost way to reduce power plant emissions. Just running existing gas-fired plants longer, and inefficient old coal-fired plants less often, can reduce emissions substantially. Replacing those obsolete coal plants with high efficiency, gas-fired, combined-cycle plants will provide a market for the abundant shale gas deposits found in many parts of the country.

The most practical and efficient policy approach to cut back carbon emissions across the entire economy is through market mechanisms, such as a cap and trade system, like those that have proven to be cost-effective in controlling other air pollutants. Putting a price on carbon will provide powerful incentives for investment in renewable energy, for improved efficiency and for bringing other low carbon options to market. We also need energy efficiency standards, reform of utility regulation and strategic infrastructure investments.

A similar agenda can be pursued globally. Readily available building blocks for developing countries include actions to encourage energy efficiency, deploy renewable energy, and avoid emissions from deforestation and land use change. These are win–win options that can promote economic growth and sustainable development while reducing emissions immediately. In the meantime, developed countries should provide assistance to help the least developed and most vulnerable countries adapt to climate changes that are already under way. These nations, already challenged, must now cope with floods, tropical storms, drought, disease, water shortages and other consequences of climate change. Developed countries must also expand public and private financial flows to fund the global energy transition, in part by reforming the Clean Development Mechanism. Limited public funding should be used strategically to catalyze much larger flows of private investment.

America's Climate Problem: The Way Forward develops these themes in greater depth. Robert Repetto, a Senior Fellow at the United Nations Foundation, untangles the complexity of today's energy and climate policy choices with admirable clarity and understanding. This latest book provides clear and sensible guidance to today's climate issues and should be read carefully by all those concerned with solving America's climate problem.

Timothy E. Wirth is President of the United Nations Foundation. Previously he represented Colorado in the US House and Senate, and served as Undersecretary of State for Global Affairs.

Acknowledgements

The author is grateful to Nan B. Burroughs for her able editorial assistance, to Robert Easton for his collaboration on the economics of adaptation, to John O. Fox for helpful comments, and to Reid Detchon, Executive Director of the United Nations Foundation, for support in the writing of this book.

Chapter 1
What We're Up Against

Climate change is raising the risks of potentially disastrous impacts and the world is at the brink of losing any realistic chance of avoiding them. The US and the rest of the world face an environmental challenge more far-reaching and more critical than any other in the course of civilization. During this century we may well irretrievably change our world for the worse. This is the unavoidable implication of an enormous scientific research effort.

Thousands of scientific studies in recent decades have examined various aspects of climate change. These studies have led to much-improved data and knowledge, and similar studies continue to be produced. This scientific literature has been assessed and summarized many times, both by the international scientific body set up to do so – the Intergovernmental Panel on Climate Change[1] – and also by national scientific bodies, including the National Academy of Science, America's most respected scientific group. Over time, as more knowledge has been gained, these assessments have become firmer and more urgent. In its most recent assessment, the National Academy stated unequivocally:

> Climate change is occurring, is caused largely by human activities, and poses significant risks for – and in many cases is already affecting – a broad range of human and natural systems. This conclusion is based on a substantial array of scientific evidence, including recent work, and is consistent with the conclusions of recent assessments by the US Global Change Research Program, the Intergovernmental Panel on Climate Change's Fourth Assessment Report, and other

assessments of the state of scientific knowledge on climate change. (National Research Council, 2010, p28)

There are still those who try to discredit and cast doubt on the ominous conclusions reached by such assessments. There are many more people who still don't take them seriously or have not paid attention to the accumulating evidence. Unfortunately, many of our congressional policymakers and supposed leaders are in these groups.[2]

Nonetheless, once the year-to-year variability in weather is averaged out, it is clear that the climate is changing. The average surface temperature around the globe increased by 0.8 degrees centigrade, or 1.5 degrees Fahrenheit, over the last century and a half, and the rate of increase quickened in recent decades, which included the warmest decades and warmest years ever recorded. Fourteen of the fifteen hottest years on record occurred between 1995 and 2009, and 2010 is on track to be the hottest year in history. Also, consistent with what we know about climate change, temperatures are rising twice as fast in the high latitudes than further south in our hemisphere.

The signs of a changing climate are all around us. Snow and ice now cover less of the globe in winter and disappear earlier in the spring. Glaciers all around the world are retreating. Sea ice is disappearing from the Arctic at a rate of 3–4 per cent per year. The oceans act as a giant thermometer, because water expands slightly when heated, and sea levels have risen by seven inches over the past century, mainly because the temperature in the top 700 meters of the oceans has warmed significantly. The oceans have absorbed most of the additional heat produced so far by the greenhouse effect. Just in the past 20 years, the heat energy absorbed in the oceans, about 0.6 watts per square meter of surface area, is as much as contained in 2 billion of the Hiroshima atomic bombs. Unfortunately, much of this heat will gradually be released back into the atmosphere, ensuring that whatever warming takes place will continue for centuries. Even Mother Nature is trying to tell us. Many plants and animals have shifted their ranges into cooler regions. Some have moved greater distances in the past 30 years than in the preceding 20,000 years (National Research Council, 2010, p39).

Climate change is driven by accumulating concentrations of heat-trapping gases in the atmosphere, the principal ones being water vapor, carbon dioxide, methane, nitrous oxide and certain fluorinated industrial gases. There is no other viable explanation. Again, as stated by the National Academy's Research Council:

> Global warming can be attributed to human activities. Many lines of evidence support the conclusion that most of the observed warming since the start of the 20th century, and especially the last several decades, can be attributed to human activities, including the following: 1) Earth's surface temperature has clearly risen over the past 100 years, at the same time that human activities have resulted in sharp increases in CO_2 and other greenhouse gases (GHGs); 2) Both the basic physics of the greenhouse effect and more detailed calculations dictate that increases in atmospheric GHGs should lead to warming of Earth's surface and lower atmosphere; 3) The vertical pattern of observed warming – with warming in the bottommost layers and cooling immediately above – is consistent with warming caused by GHG increases and inconsistent with other causes; 4) detailed simulations with state-of-the-art computer models are only able to reproduce the observed warming trend and pattern when human-induced GHG emissions are included; 5) The observed warming, especially the warming since the late 1970s, cannot be attributed to natural variations. (National Research Council, 2010, p30)

Atmospheric concentrations of greenhouse gases are already higher than at any time in the previous 800,000 years. Concentrations are rising rapidly because the sources of greenhouses gases in the atmosphere, mainly emissions from fossil fuel combustion and from land use changes, exceed the sinks for greenhouse gases in the oceans, soil and vegetation.

The difference accumulates in the atmosphere, where carbon dioxide stays for centuries. Immediate stabilization of concentrations would require an immediate 50 per cent reduction in emissions. Any delay in making these reductions implies that concentrations will continue to rise.

Unfortunately, emissions are rising, not falling, so the increase in atmospheric greenhouse gas concentrations is accelerating. The emissions growth rate since 2000 has been faster than for the most fossil fuel-intensive emissions scenarios developed by the Intergovernmental Panel on Climate Change in the late 1990s. Global emissions growth since 2000 has accelerated because declining trends in the energy intensity of economic output and the carbon intensity of energy used were reversed. The recession of 2007–2009 meant only a slight temporary break in these developments. The growth rate in emissions is highest in rapidly developing economies such as China. Together, lower and middle income countries, accounting for 80 per cent of the world's population, were responsible for 73 per cent of global emissions growth in 2004 but only 41 per cent of global emissions (Raupach et al, 2007). Emissions are also still rising in the US, the world's second largest emitter.

Projections issued in mid-2010 by the Energy Information Agency predict that without strong new policies global emissions would rise by an additional 43 per cent by 2035. Most of the increases would come from the rapidly growing developing countries, but emissions from the US and other industrialized countries would also increase. If these trends are not reversed and emissions sharply reduced, atmospheric concentrations could reach 800 to 1500 parts per million by the end of the century, three to five times the pre-industrial level, with potentially devastating consequences.

INCREASING RISKS

How strongly these rising concentrations will effect climate in the long run is uncertain. For example, the best estimate of the long run impact of a doubling of CO_2 concentrations above pre-industrial levels, from 285 to 570 parts per million,[3] is a rise of 3 degrees centigrade (5.4 degrees Fahrenheit) in global average temperatures, but with significant probability it could be as low as 2 or as high as 4.5 degrees centigrade (3.6 or 8 degrees Fahrenheit). The uncertainty stems mainly from feedbacks in the climate system: as Earth's surface and lower atmosphere warms, more moisture evaporates from the earth and ocean surfaces and the atmosphere becomes more humid. Since water vapor is also a strong greenhouse gas, it amplifies the effect of carbon dioxide emissions (Sokolov et al, 2009).

In addition, as the ocean warms, its ability to absorb even more heat diminishes, and as it becomes more saturated with carbon dioxide, its ability to absorb even more CO_2 also declines. Similarly, over time, forests and other plants become less able to take up more CO_2, which weakens another important sink for carbon in the global cycle. Furthermore, computerized climate models predict that warming will eventually begin to limit photosynthesis and to accelerate the decomposition of organic matter in soil, reducing the capacity of terrestrial ecosystems to store carbon. Warming will also result in slower vertical ocean circulation, reducing the amount of carbon that sinks from the shallow waters to stay in the deep ocean (Friedlingstein, 2008). The pace and magnitude of these secondary effects are uncertain, so tampering with the global climate is risky.

The risks are increased because we can't fully detect what we have already done and will not be able to tell for decades. The climate responds gradually, with a lag, to changes in greenhouse gas concentrations as the oceans absorb heat and gradually reach a new equilibrium with the atmosphere. The changes in climate we're now experiencing are largely the consequences of the increases in GHG concentrations some decades ago. Managing the climate problem is like driving a car with a brake that works only some time after the pedal is depressed and that can slow the car down only gradually. With brakes like that, the driver would be well advised to stay very clear of danger.

Many of the impacts of a changing climate are already being felt, though most people have trouble connecting these manifestations to underlying trends. The US Global Change Research Program has documented many of these impacts (US Global Change Research Program, 2009). Not all of the effects are troublesome, of course: warmer temperatures in winter mean fewer cold snaps and lower heating bills. Somewhat warmer and wetter weather and higher atmospheric CO_2 concentrations in the farm belt may mean higher yields, at least in the short run. But many of the effects are damaging and will become increasingly so as climate change continues. Within the US, effects will vary regionally. At the extreme, in Alaska, where change has been twice as rapid as in the lower 48 States, the retreat of sea ice has already left many coastal communities exposed to shoreline erosion in winter storms, forcing entire villages to move. Melting permafrost has destroyed buildings and infrastructure. These impacts will intensify.

In general, the drier regions will become even drier, the wetter regions even wetter. Overall precipitation is already up 5 per cent over the past several decades and the percentage of precipitation that has fallen in the most extreme storms is up 20 per cent, leading to increased flooding along the river basins. In the mountain and western States more of the precipitation is falling as rain, not snow, and the snow pack that feeds the hydropower and irrigation systems over the summer months is melting earlier, reducing water supplies in late summer and fall. In the Southeast, the South and the West, higher temperatures and drier weather are exacerbating the frequency and extent of forest and grass fires, endangering or destroying homes and properties. Throughout the South, the Southwest and the Midwest, summer heatwaves are becoming more frequent and intense, making cities like Phoenix and Houston very unattractive. In these regions, droughts will become more frequent and long-lasting. As winter temperatures rise, more pest populations are expanding their range and wintering over, leading to agricultural and forest losses. Infectious diseases will also spread, including tropical fevers such as Dengue and malaria that have rarely appeared in the US. Warmer, more humid weather will encourage formation of smog and atmospheric particulates, degrading air quality and aggravating asthma and other bronchial problems (Epstein, 2005).

As the sea level and ocean temperatures continue to increase, there will be more coastal flooding and beach erosion, especially during storm surges. Low-lying areas will be threatened with inundation, especially in Florida, the South Atlantic and the Gulf Coast. The most intense Category 4 and 5 hurricanes will strike more frequently, with tremendous damage to densely populated coastal cities. As Chapter 5 explains, an intense hurricane striking the New York City region would probably cause hundreds of billions of dollars in damage, and the likelihood of such a disaster is steadily increasing.

As time has passed, scientists who have studied the issue have found more reasons for concern for the future (Smith et al, 2009). Weather records show that these extremes are happening more often. The number of exceptionally hot days during the year, including multi-day hot spells and heatwaves, has been increasing over the past several decades. Drought conditions have become more frequent, along with higher temperatures, in the Southwest and southern Great Plains. More of the annual precipitation across the US has been falling in heavy storms. For example, in the upper Midwest, there has been a 50 per cent increase over the last century in the percentage of days with precipitation exceeding four inches, exacerbating flood risks, which trail only hurricanes as the cause of weather damages. The intensity of hurricanes in the Atlantic and the occurrence of the more extreme hurricanes have increased substantially over the past four decades as the result of rising sea-surface temperatures. Already the number of the most intense hurricanes making landfall has increased by almost 50 per cent in the last 15 years, compared to the previous period (US Climate Change Science Program, 2008).

The acidity of the oceans has already increased by 30 per cent since pre-industrial times, because they have absorbed about a third of all carbon emitted and converted it to carbonic acid. As this continues, it impedes shell formation in marine organisms, including larvae of commercial species and other organisms at the base of the aquatic food chain. The potential impact on our ocean fisheries and on Hawaii's valuable coral reefs is severe.

Some years ago, many people thought that an increase of three or four degrees in average temperatures could be dealt with easily. After all, people spend most of their time indoors in their houses, offices and shopping malls. Tell that to the people in Louisiana, Tennessee, Minnesota and Rhode Island whose homes, schools and businesses were flooded out, or to the people whose homes were burnt by wildfires in California. It is now better understood that as average weather conditions change, the frequency of extreme weather conditions change disproportionately, and it is those extreme weather conditions that do by far the most damage. More frequent floods, intense hurricanes, tornados, heatwaves and droughts are the most risky consequences of climate change. According to the chief of climate analysis at the National Climatic Data Center, 'The climate is changing. Extreme events are occurring with greater frequency and in many cases with greater intensity' (Gillis 2010).

Truly, though the risks to the US are severe, the risks to less developed parts of the world are potentially disastrous. Many of these countries are already struggling with rising populations, overstressed natural resources, limited economic capacities and inadequate governance. Climate change will introduce additional stresses to already difficult or desperate conditions. Low-lying densely populated coastal countries such as Bangladesh will lose substantial fractions of their land to inundation and salinity and devastating typhoons will roll inland for scores of miles. Semi-arid subtropical areas in Mexico, Africa, the Mediterranean and Central Asia will become hotter and drier, reducing output of staple food crops of corn, wheat and rice, and threatening those populations with scarcity or worse. Water supplies in countries that are already water-short are likely to become even more inadequate while water needs for irrigation, power and urban uses continue to increase. Regions subject to flooding, such as Eastern India and Southeast Asia, will see even more flooding. Countries that are even now unable to provide adequate health services to their populations will face an increasing burden of infectious diseases such as malaria, diarrhea and respiratory diseases. Natural disasters will lead to more human suffering, refugee migration, and political and social instability.

The World Bank devoted its 2010 World Development Report to the challenge of climate change in the developing world. Its in-depth assessment concluded that climate change likely in this century, if drastic action is not taken, will be devastating for developing countries, which are more reliant on natural resources and more exposed to damages. For example:

> The possible dieback of the Amazon rain forest; complete loss of glaciers in the Andes and the Himalayas; and rapid ocean acidification leading to disruption of marine ecosystems and death of coral reefs. The speed and magnitude of change could condemn more than 50 per cent of species to extinction. Sea levels could rise by one metre this century, threatening more than 60 million people. ... Agricultural productivity would likely decline throughout the world, particularly in the tropics, even with changes in farming practices. ... Between 100 million and 400 million more people could be at risk of hunger, and 1 billion to 2 billion more people may not have enough water to meet their needs. (World Bank, 2010, pp4–5)

Some Americans might say that what happens in the developing world means little to us, ignoring our current involvements in such resource-stressed countries as Yemen, Somalia and Afghanistan and our problems with illegal migrants from Mexico and Central America coming north in search of economic opportunities. America's national security agencies would disagree. They have identified climate change as a significant security threat in this century (Broder, 2009). Those who are unconvinced should repeat to themselves one word: Pakistan. Pakistan is a country with weak democratic traditions, a weak civilian government, and a powerful nuclear-armed military establishment obsessed with its territorial conflict with neighboring India, which has led to three wars since independence. It has been the source of nuclear technology sold to Iran and North Korea. Pakistan harbors leaders of the Taliban and Al Qaeda and continues to be the launching pad for terrorist attacks in India, Pakistan itself, Afghanistan and the US. It is close now to becoming a failed state, facing internal insurgencies in the Frontier Agencies, Baluchistan and Sindh in the south, as well as conflicts with militant Islamic groups throughout the country.

Population growth rates and birth rates are still high, because education, especially of women, has been neglected. Agriculture still contributes one-quarter of total income and employs one-half of the labor force. Farming is almost totally dependent on irrigation based on runoff from the Himalayas that feeds the canals and recharges the aquifers. Water supplies are already inadequate: in many areas surface flows cannot replace the irrigation water being pumped out, so aquifers are being depleted.

How will climate change affect Pakistan, a mostly arid country? Climate change is melting the Himalayan glaciers and more precipitation in the mountains is falling as rain and running off. Initially, the glacial melt and intensification of the monsoon will increase runoff, but earlier in the dry season when irrigation is less critical. Pakistan lacks the storage capacity to capture this runoff, and existing reservoirs are rapidly silting up, so damaging flooding like that in 2010 is increasingly likely. Over time, flow from shrunken glaciers will decline and water availability in Pakistan will diminish by 30 to 40 per cent (Briscoe and Qamar, 2006).

Declining water availability in the face of rising water demands is likely to cripple Pakistan's agriculture and exacerbate internal and external conflicts. One monumental agreement that India and Pakistan achieved after their partition was the Indus Water Treaty, dividing the waters of the Indus Basin between the two countries. It has survived despite three wars. Will it survive a 30 per cent decline in water availability in the shared Indus Basin or will there be heightened conflict between these two nuclear powers? For that matter, will internal conflicts over water between the upstream province of the Punjab and the downstream provinces of Sindh and Baluchistan lead to even greater internal instability? Will the growing number of under-educated, under-employed, radicalized young Muslim men with no employment opportunities in the countryside provide ready recruits for domestic and international militants and terrorists? Is it possible to believe that these issues have no importance for the US?

Increasing concentrations and degrees of warming imply escalating risks, a point emphasized by the influential Stern Review on the Economics of Climate Change, an assessment carried out for the UK Department of the Treasury (Stern, 2007). Damages from climate change rise nonlinearly at an increasing rate as temperature warms, the frequency of extreme events increases and ecosystems are pushed beyond their ranges of tolerance. The risks of an additional increase in global temperatures from three to four degrees are much greater than those from two to three degrees and those in turn even more than those from one to two degrees of warming. For example, the atmosphere's capacity to hold moisture increases exponentially as temperatures rise, implying an intensification of the hydrological cycle and more extreme storms. On top of that, damages rise exponentially with height of flood crests and the wind speeds attained by such storms. Together, these imply that the added risks of storm damage increase very rapidly with more warming.

THE SCARIEST PARTS

The most disturbing realization is that climate change could spiral out of control, making it impossible to stop even with sharp reductions in emissions. In several different ways, higher temperatures unleash geological and biological forces that lead to even faster warming, creating a vicious circle of climate change. These mechanisms are not just theoretical; several are already observable.

The melting of sea ice eliminates the cover over the Arctic oceans, exposing those waters to warming from the sun and leading to even more melting. The loss of ice replaces a shiny white surface that reflects back the sun's radiation with a dark surface that absorbs it. The mutually reinforcing disappearance of sea ice and warming of Arctic oceans contribute to the rapid warming of the high latitudes (Soden and Held, 2006).

This warming unleashes an even more powerful amplifying mechanism. Enormous amounts of carbon are stored in beds of frozen permafrost across the Canadian and Siberian Arctic. When this permafrost melts, carbon dioxide and methane are released from the previously frozen and anaerobic soils. Methane is a greenhouse gas 20 times more powerful than carbon dioxide, though shorter lived in the atmosphere. Permafrost melting and these greenhouse gas releases are already being observed in Alaska, Russia and elsewhere during the Arctic summers at faster rates than previously estimated (Dorrepaal, 2009). These releases accelerate warming, diminishing snow and ice cover, and more permafrost releases in a dangerous vicious circle.

An analogous feedback mechanism extends underwater into the Continental Shelf, where vast quantities of methane in the form of methane hydrates are confined on the ocean floor by low temperatures and high water pressures. Studies in the East Siberian Arctic Shelf, a shallow area of 2 million square kilometers, have found that the permafrost is leaking and methane is bubbling up through the water into the atmosphere. Scientists estimate that about 8 million tons of methane are now being released each year, an amount equal to that released by all the rest of the world's oceans. Escape of even one per cent of the methane previously confined in the Shelf could trigger abrupt, runaway climate change. According to the lead author of the study, 'Our concern is that the subsea permafrost has been showing signs of destabilization already. (Shakhova et al, 2010). If this process is well under way, it may soon become impossible to stop it.

There are also strong feedbacks on land arising from the effect of hotter drier conditions in some regions on forest fires, which not only release carbon directly into the atmosphere but expose previously covered soil to the loss of stored soil carbon. Forest fires globally already consume an area as large as Texas every year and account for almost half as much carbon emissions as those from fossil fuel combustion. Forest fires also disperse large quantities of black carbon soot, which absorbs heat. Large forested areas such as central Russia and the Amazon could experience even more widespread and frequent fires, which would add to the warming and create the conditions for even more fires.

These reinforcing mechanisms are matters of the most serious concern. They could well lead during this century to climate change spiraling out of human control. One potential extreme consequence could be abrupt climate change lasting centuries or millennia. Though this might seem to be a fanciful doomsday scenario, there is no reason to think that Earth's climate is inherently stable. On the contrary, the fluctuation between warm and icy eras in geological time is well known. Less well known is the fact that in the past Earth's climate has changed dramatically, by 8 to 10 degrees centigrade, in a matter of decades. Though scientists consider the possibility of such abrupt changes unlikely, they don't rule it out (National Research Council, 2002). The most recent National Academy assessment reported thus:

> Paleoclimate records indicate that the climate system can experience abrupt changes in as little as a decade. The Earth's temperature is now demonstrably higher than it has been for several hundred years, and greenhouse gas concentrations are now higher than they have been in at least 800,000 years. These sharp departures from past and recent climate raise the possibility that tipping points or thresholds for stability may be crossed as the climate system warms, leading to rapid or abrupt changes in climate. (National Research Council, 2010, p30)

Fortunately, not all of these feedback mechanisms reinforce warming trends. Increased cloud formation in a more humid climate may reduce the rate of warming by reflecting solar radiation back away from Earth if the clouds are dense, low level cumulus clouds rather than thin, high level cirrus clouds that are less effective in reflecting solar radiation but more effective in absorbing outgoing infrared radiation. Scientists are still not certain about the net effect of increased cloudiness but agree that it is an important aspect of climate change. Further complicating the issue, low level cloud formation is increased by aerosols released by fossil fuel combustion. These tiny particulates form atmospheric brown clouds that travel long distances and form nuclei around which water vapor droplets can form. Fossil fuel combustion, especially in the rapidly industrializing middle income countries, has exacerbated air pollution but has suppressed much of the warming from greenhouse gas emissions by increasing haze and cloudiness. Globally, atmospheric brown clouds may so far have masked as much 47 per cent of the global warming by greenhouse gases. Ironically, the shift to cleaner fuels and efforts to reduce air pollution as living standards rise in rapidly industrializing countries will probably reverse this process, making climate stabilization even more difficult (Ramanathan and Feng, 2009).

WHAT MUST BE DONE

To forestall even greater risks from climate change than those that have already been created, greenhouse gas emissions must be drastically reduced to a level not exceeding the available terrestrial and oceanic sinks, taking into account that those sinks are themselves diminishing as concentrations rise. A reduction of 50 to 70 per cent below current levels is essential.

What matters most is the total accumulation of long-lived greenhouse gases in the atmosphere. Therefore, cumulative global emissions over this century will largely determine the extent of climate change. For example, to have a good chance of limiting global warming to two degrees centigrade above pre-industrial levels, which UNFCCC signatories agree is the maximum change that might avoid dangerous impacts, cumulative global emissions of carbon dioxide equivalents should not exceed one trillion tons (Allen et al, 2009). Half of that amount has already been emitted during the industrial era. Today, with rapid growth in population and the size of the global economy, emitting only that same amount from now on necessitates rapid transitions away from fossil fuel combustion and deforestation.

Because of considerable uncertainties about the feedbacks and climate sensitivities discussed above, it's impossible to know conclusively that limiting future emissions to that extent will actually confine warming below dangerous levels. Table 1.1 below reproduces probability estimates from a suite of climate models linking atmospheric concentrations of carbon dioxide to eventual warming (Stern, 2007, Chapter 8). It shows that limiting concentrations to 450 parts per million, a very ambitious target, would have only about a 50 per cent chance of holding warming to two degrees centigrade. Higher concentrations would eliminate any realistic chance of doing so and would almost guarantee that warming will continue into levels of rapidly rising risks.

Table 1.1 Probabilities of warming at various CO_2e levels

Concentration (CO_2e)	Probability of warming 2°C	Probability of warming 3°C	Probability of warming 4°C
450ppm	38–78%	6–18%	1–3%
500ppm	61–96%	18–44%	4–11%
550ppm	77–99%	32–69%	9–24%
650ppm	92–100%	57–94%	25–58%
750ppm	97–100%	74–99%	41–82%

Source: Stern (2007)

There are various trajectories consistent with the same overall accumulations of greenhouse gases. The longer the delay in beginning to reduce emissions, the faster the subsequent rate of decline would have to be to achieve the same cumulative emissions total. For example, if global emissions were to start falling immediately, then to limit ultimate accumulations to 500 parts per million, they would have to fall over the next half-century by an average of 3 per cent per year. If the decline were delayed until 2020, then the rate of decline would have to accelerate to 4–6 per cent per year to achieve the same objective. However, to achieve the limit of 450 parts per million, the limit most consistent with avoiding dangerous impacts, there are no degrees of freedom. Reductions in global emissions must start almost immediately and continue at an annual rate of 7 per cent per year. This trajectory is very likely unattainable, since emissions are still rising in almost all countries and there is as yet no global agreement to stop the increase. This discouraging conclusion implies that the world is probably already locked into a dangerously risky degree of climate change.

One conclusion that might be drawn from these analyses is that a special effort should begin immediately to reduce emissions of extra-powerful, relatively short-lived greenhouse gases such as methane, which is at least 20 times as powerful as carbon dioxide, and some fluorinated industrial gases that are thousands of times more powerful, because doing so would greatly help in the short run to reduce the growth rate of concentrations. Some of the options for reducing these emissions, such as stopping the flaring of natural gas from oil wells and reducing leakage from natural gas pipelines, are very cost-effective and might even save money. Some misguided critics of international programs criticized activities in China and India to reward reductions in the industrial gases because doing so was 'too profitable' and didn't warrant such rewards. Just the opposite is true: all the highly cost-effective, easily implemented short term measures should be exploited as soon as possible.

What do these trajectories of global emissions reductions imply for the US? Understandably, the apportionment of responsibility for reducing emissions has been a contentious subject of international negotiations and will continue to be so. No comprehensive international agreement has yet been reached. Two aspects are crucial here: first, because the US – by virtue of its past emissions – has been by far the largest contributor to the stock of greenhouse gases now in the atmosphere, because the US is the wealthiest of all the major emitting countries and because per capita emissions in the US are far above the world average, it is inconceivable that any international agreement can be reached if the US is to make a smaller percentage reduction than other countries. On the contrary, opinion is almost universal in other countries that the US must do considerably more. Europe has evidently already accepted such a responsibility and the climate bills recently introduced in the US Congress have adopted a 2050 emissions reduction target of 83 per cent below 2005 levels, consistent with this obligation.

Second, there is an important difference between the responsibility for emissions reduction and the implementation of such reductions within the US. Largely because of American insistence, the Kyoto Protocol created mechanisms whereby countries could pay for emissions reductions implemented in other countries, taking advantage of more cost-effective opportunities to be found abroad. Ideally, emissions reductions will take place wherever they can most cost-effectively be carried out, no matter who pays for them. Ironically, though the US never ratified the international agreement it worked so hard to shape, those mechanisms – Joint Implementation and the Clean Development Mechanism – have been in operation since 2008 and have provided proof of concept. Financing emissions reductions in other countries, especially developing countries, can be a cost-effective way of supplementing emissions reductions carried out domestically. Though the existing mechanisms are imperfect and can be significantly improved, as Chapter 6 explains, the principle is sound.

Reducing emissions by more than 80 per cent over a 40-year period will require a radical overhaul of America's energy systems and patterns of energy use. This cannot be accomplished without tackling the major emissions sources. Table 1.2 below shows what those are.

What this inventory makes clear is that carbon dioxide is by far the largest source of anthropogenic greenhouse gases, amounting to 85 per cent of emissions and staying in the atmosphere for very long periods of time. Moreover, fossil fuel combustion is by far the largest source of carbon dioxide emissions. Unless it proves possible to capture and permanently sequester out of the atmosphere the enormous quantities of carbon dioxide produced by burning fossil fuels, emissions reductions on the scale required implies a radical move away from fossil fuels as the primary energy source.

No less clear is the conclusion that the energy basis for transportation systems and for electricity generation must be transformed. These two systems account for the majority of US greenhouse gas emissions. Exhaust emissions from more than 160 million vehicles obviously cannot be captured and sequestered, so the only available strategies are electrification of transportation or conversion to renewable energy fuels. Both strategies are feasible and must come into play. In order to reduce emissions from electricity generation drastically, there are also

Table 1.2 Sources of US greenhouse gas emissions, 2008

	Million tons, CO_2e	% of total
Total GHG emissions	6958	100
Total CO_2 emissions, of which	**5921**	**85**
Electricity	2364	34
Transportation	1785	26
Industrial	819	12
Residential	343	5
Commercial	220	3
Non-fuel industrial	348	5
Total methane emissions, of which	**568**	**8**
Enteric fermentation	141	
Landfills	126	
Natural gas systems	96	
Coal-bed methane	68	
Manure management	45	
Total N_2O emissions, of which	**318**	**5**
Soil and manure management	233	
Total industrial gases	**150**	**2**

Source: US Environmental Protection Agency (2010)

just two available strategies. One, already mentioned, is capturing and sequestering carbon dioxide on a very large scale, but the feasibility of this strategy has not yet been demonstrated. The other is a shift from fossil fuels, especially coal, as the primary energy source for electricity generation to non-carbon sources, including nuclear, hydro, wind, solar and tidal energy.

Cutting across these strategies, an essential element of a successful transition must be a very large improvement in energy efficiency –

obtaining more useful energy services from each unit of energy used. Fortunately, the opportunities for doing this are large and widespread. Table 1.3 disaggregates total carbon dioxide emissions in terms of the end-use sectors responsible. Looking at the data this way again directs attention to transportation and shows by how little electricity has as yet penetrated that sector. It also shows the importance of industrial, residential and commercial uses, which use by far the greatest shares of all electricity generated. Industrial emissions stem significantly from both combustion and from electricity use. Synergies between the two can be exploited by taking advantage of many available opportunities to use waste heat from industrial processes to generate electricity through cogeneration. Electricity use has increased particularly rapidly in the residential sector because of the increasing number of home computers, printers, televisions and other electronic items in people's houses. Improving end-use efficiency in these sectors can not only reverse the growth in US emissions but also contribute significantly to their reduction.

The necessary reductions in greenhouse gas emissions will not occur spontaneously without strong policy inducements. Relying on voluntary measures by firms and households, the stance adopted by the administration of George W. Bush, was ineffectual and allowed emissions to increase. During the past decade, despite lack of leadership from the White House, Congress has enacted some policy measures. They have slowed the rate of increase but have not been enough to bring about a decline, let alone the rapid reductions that are required.

The Energy Policy Act of 2005 put in place new tax incentives to encourage energy conservation. Tax credits were enacted for outlays to make new or existing commercial and residential buildings more energy-efficient. Credits were also provided to businesses and households for purchases of solar equipment and to manufacturers of more energy-efficient appliances. Congress also enacted tax credits for purchasers of hybrid, fuel cell or other lean-burn, fuel-efficient vehicles;

Table 1.3 US carbon dioxide emissions by end-use sector, 2008

	Million tons, CO_2e
Total emissions	5572
Transportation sector	**1789**
Combustion	1784
Electricity	5
Industrial sector	**1511**
Combustion	819
Electricity	692
Residential	**1185**
Combustion	343
Electricity	842
Commercial	**1045**
Combustion	220
Electricity	825

Source: US Environmental Protection Agency (2010)

for the production of biodiesel and ethanol; and for the construction of fueling stations for alternative fuels. Many of the above tax provisions, however, were limited in extent, duration or eligibility (Congressional Research Service, 2006).

In the same legislation, along with these inducements to reduce energy demand, Congress provided new tax incentives to increase energy supply, not only of non-fossil fuels but also of traditional fossil fuels. In the electricity sector, credits were enacted for renewable energy investments and production, for nuclear generation, for new investments in transmission lines, but also for advanced coal-based electric power plants. In addition, more tax breaks were provided for the production, transportation, refining and distribution of oil and natural gas and for the production of coke and coke gas from coal. The net effect of this legislation would be difficult to assess but clearly showed the continuing political influence of traditional energy industries and fell far short of the strong climate policy measures that are needed.

In 2007 Congress tried again with the Energy Independence and Security Act (EISA) (Congressional Research Service, 2007). This legislation enacted more extensive provisions to reduce greenhouse gas emissions but, as before, Congress deliberately refrained from labeling it a climate bill. The bill tightened Corporate Average Fuel Economy (CAFE) standards for cars and light trucks to 35 miles per gallon. New vehicles are to achieve this by 2020, though many models sold in 2007 already met or exceeded this level. Such standards are a relatively weak policy instrument, compared to gasoline taxes, because they apply only to sales of new cars, about 10 million vehicles per year, not to the approximately 160 million vehicles on the road. Moreover, there is a 'rebound' effect, in that lower fuel consumption lowers vehicle operating costs and encourages owners to drive more miles.

The bill also imposed a Renewable Fuel Standard requiring an increase in non-fossil transportation fuels from 9 billion gallons in 2008 to 36 billion gallons in 2022, all of the increase to come from cellulosic ethanol and other advanced biofuels rather than from corn ethanol, which does little to reduce GHG emissions. Current gasoline consumption is approximately 140 billion gallons per year. By 2022 about 25 per cent of fuel consumption would rely on renewables, compared to the 80–85 per cent achieved right now in Brazil, using ethanol from sugarcane. Also, fiscal support was strengthened for R&D to improve technologies for batteries and hybrids and other fuel-efficient vehicles. Federal agencies were required to market support for fuel-efficient vehicles through purchases and a general requirement to reduce petroleum use by 20 per cent.

The EISA contained many provisions to encourage greater energy efficiency. Minimum efficiency standards were adopted by statute for lighting and a wide range of appliances, such as refrigerators and freezers, washers and dryers, dishwashers and air conditioners. Such standards nudge consumers toward purchasing decisions that are cost-saving over the life of the appliance. Funding was also increased for home weatherization programs and a provision was inserted requiring that federal agencies make a major effort to reduce energy use in government buildings by 30 per cent over a decade.

The legislation included some significant elements of a national climate policy but is also significant for what it did not include. First and foremost, it did not include any version of a national cap and trade system for GHG emissions, nor any carbon tax, although cap and trade systems had already been adopted in Europe and in the Northeastern States' Regional Greenhouse Gas Initiative. Congress declined to impose a price on carbon emissions, an essential core climate policy. Neither did it enact a national Renewable Portfolio Standard (RPS) that would require electric utilities to generate some percentage of their power from renewable resources, though almost half the States had some sort of RPS by 2007. Further, the final version of the bill failed to repeal approximately $22 billion in annual subsidies to the oil and gas industries, a revenue-raising measure that was included in the bill the House passed as a way to pay for a four-year extension of the production tax credit for renewable power generation. The George W. Bush White House had threatened to veto the bill if these subsidies were repealed (Congressional Research Service, 2007).

The Obama administration has adopted a more proactive stance on America's climate problem and included several additional measures in the 2009 stimulus package, the American Recovery and Reinvestment Act, including more funding for efficiency upgrades in private and public buildings, for build-out of the 'smart' electricity grid and for mass transit. It has supported legislative proposals for a national system that would effectively put a price on carbon emission. This initiative is discussed in Chapter 3. First, however, the following chapter lays out a feasible energy transition based on known, cost-effective technologies that can achieve the required reduction in greenhouse gas emissions while strengthening the economy and improving the environment and national security.

NOTES

1 In 2009, opponents of actions to reduce carbon emissions seized on a few errors in the International Panel on Climate Change's Fourth Assessment Report to try to discredit the entire report, the result of a six-year effort by more than 1000 scientists from 150 countries. In March 2010, an open letter signed by 250 US scientists, mostly from universities and research institutes, wrote that 'none of the handful of mis-statements (out of hundreds and hundreds of unchallenged statements) remotely undermines the conclusion that "warming of the climate system is unequivocal" and that most of the observed increase in global average temperature since the mid-20th century is very likely due to observed increase in anthropogenic greenhouse gas concentrations'; see www.openletterfromscientists.com, accessed March 2010.

2 The Obama administration, however, recognizes the urgency of the climate challenge. In its most recent report to the United Nations Framework Convention on Climate Change, the US Department of State wrote that 'global warming is unequivocal and primarily human-induced'; see http://unfccc.int/resource/docs/natc/usa_nc5.pdf, accessed 21 April 2010.

3 Concentrations at the time of writing are 387 ppm.

REFERENCES

Allen, M., Frame, D., Huntingford, C., Jones, C., Lowe, J., Meinshausen, M. and Meinshausen, J. (2009) 'Warming caused by cumulative carbon emissions towards the trillionth tonne', *Nature*, vol 458, pp1163–1166, www.nature.com/nature/journal/v458/n7242/full/nature08019.html

Briscoe, J. and Qamar, U. (2006) *Pakistan's Water Economy: Running Dry*, Oxford University Press for the World Bank, Karachi, Pakistan

Broder, J. (2009) 'Climate change seen as threat to national security', *New York Times*, 8 August, pA1

Congressional Research Service (2006) *Energy Policy Act of 2005, Summary and Analysis of Enacted Provisions*, Washington, DC

Congressional Research Service (2007) *The Energy Independence and Security Act of 2007, Summary of Major Provisions*, Washington, DC

Dorrepaal, E., Toet, S., van Logtestijn, R., Swart, E., van de Weg, M., Callaghan, T. and Aerts, R. (2009) 'Carbon respiration from subsurface peat accelerated by climate warming in the Subarctic', *Nature*, vol 460, pp616–619, www.nature.com/nature/journal/v460/n7255/index.html#af

Epstein, P. (2005) 'Climate change and human health', *New England Journal of Medicine*, vol 353, pp1433–1436

Friedlingstein, P. (2008) 'A steep road to climate stabilization', *Nature*, vol 451, pp297–298

Gillis, J. (2010) 'In weather chaos, a case for global warming', *New York Times*, 15 August

National Research Council (2002) *Abrupt Climate Change: Inevitable Surprises*, National Academies Press, Washington, DC

National Research Council (2010) *Advancing the Science of Climate Change*, National Academies Press, Washington, DC

Ramanathan, V. and Feng, Y. (2009) 'Air pollution, greenhouse gases and climate change: Global and regional perspectives', *Atmospheric Environment*, vol 43, pp37–50

Raupach, M., Marland, G., Ciais, P., Le Quere, C., Canadell, J., Klepper, G. and Field, C. (2007) 'Global and regional drivers of accelerating CO_2 emissions', *Proceedings of the National Academy of Science*, vol 104, pp10,288–10,293, www.pnas.org/

Shakhova, N., Semiletov, I., Salyuk, A., Yusupov, V., Kosmach, D. and Gustafsson, O. (2010) 'Extensive methane venting to the atmosphere from sediments of the East Siberian Arctic Shelf', *Science*, vol 327, pp1246–1250

Smith, J., Schneider, S., Oppenheimer, M., Yohe, G., Hare, W., Mastrandrea, M., Patwardhan, A., Burton, I., Corfee-Morlot, J., Magadza, C., Füssel, H,

Pittock, A., Rahman, A., Suarez, A. and van Ypersele, J. (2009) 'Assessing dangerous climate change through an update of the Intergovernmental Panel on Climate Change (IPCC) "reasons for concern"', *Proceedings of the National Academy of Science*, vol 106, pp4133–4137, www.pnas.org

Soden, B. and Held, I. (2006) 'An assessment of climate feed backs in coupled ocean–atmosphere models', *Journal of Climate*, vol 19, pp3354–3360

Sokolov, A., Stone, P., Forest, C., Prinn, R., Sarofim, M., Webster, M., Paltsev, S., Schlosser, C., Kicklighter, D., Dutkiewicz, S., Reilly, J., Wang, C., Felzer, B., Melillo, J. and Jacobyokolov, H. (2009) 'Probabilistic forecasts for 21st century climate based on uncertainties in emissions and climate parameters', *Journal of Climate*, vol 22, pp5175–5204

Stern, N. (2007) *The Stern Report on the Economics of Climate Change*, Cambridge University Press, Cambridge, UK, Chapter 3

US Climate Change Science Program (2008) 'Weather and climate extremes in a changing climate: Regions of focus – North America, Hawaii, Caribbean and US Pacific Islands', Synthesis and Assessment Product 3.3, US Department of Commerce, Washington, DC, www.globalchange.gov

US Environmental Protection Agency (2010) *Inventory of Greenhouse Gas Emissions and Sinks, 1990–2008*, Washington, DC

US Global Change Research Program (2009) *Climate Change Impacts on the United States*, Washington, DC, www.globalchange.gov/usimpacts

World Bank (2010) *World Development Report: Development and Climate Change*, Washington, DC

Chapter 2
What We Must Do: Meeting the Energy Challenge

THE ENERGY CHALLENGE

Greenhouse gas concentrations in the atmosphere must be stabilized at a level somewhere between 450 and 500ppm if these extreme risks to the US and the rest of the world from climate change are to be avoided. To achieve this, within this decade carbon dioxide emissions must begin falling by about 3 per cent per year and do so over the next 50 years. By then, emissions must be about 80 per cent below current levels. This crucial trajectory requires a sharp break with our past dependence on fossil fuels.

Breaking that dependency is also important for our national security. America now imports more than two-thirds of its oil, much of it from countries that are unstable, undemocratic or unfriendly to us. This situation limits the government's foreign policy options with respect to Iran, Iraq, Venezuela or Saudi Arabia, because their supplies to the world oil markets are irreplaceable. It also requires a large military presence in the Middle East and elsewhere to defend our energy interests.

For security reasons, the US must also create an energy system that is less brittle and vulnerable to potential terrorist attack or accidents. Large central power plants, pipelines, LNG terminals, high voltage transmission lines and other energy infrastructure are all potential targets. We need to create a more resilient, more decentralized energy system that can't be crippled if a few key components are put out of operation.

Finally, the US faces the economic challenge of restoring economic prosperity. The present energy system inevitably implies increasing long term scarcity and increasing economic burdens. Oil imports already account for much of the country's balance of payments deficit, which has depressed the exchange rate and eroded our purchasing power abroad. In 2008 the US was spending half a million dollars per minute on oil imports. In the future, these burdens will escalate because the costs of finding and producing new supplies are rising. One hundred and fifty years ago, the nascent oil industry found petroleum seeping out of the ground in Pennsylvania. Now, to replace declining reserves, the industry must drill a mile or more under the ocean or in the far north Arctic. The 2010 spill in the Gulf of Mexico off Louisiana's shores showed how risky those operations are. The costs of supplying coal from domestic mines are also rising because more and more layers of overburden must be removed by highly energy-intensive equipment to get at the deeper reserves.

The US needs a transition away from this dependency and an economic stimulus that will spark a new wave of investment, innovation and employment, creating new industries, new technologies and new job opportunities. Other countries are already embarked on this transition and will take the lead unless the US rises to the challenge. So, for example, our share of the solar industry, which was 80 per cent 25 years ago, has now fallen to under 10 per cent. Economic competitiveness in a new energy era requires a rapid transition.

Though a transformation of this scope may seem impossible to some people, energy transitions over the course of a half-century are nothing new. They have occurred repeatedly in American history (Grubler, 2008). In the mid-19th century, almost all American houses were illuminated with whale oil lamps and whaling was a huge industry until kerosene and gas lamps took over. Since that transition, New Bedford has not fully recovered but Nantucket is doing quite well.

Also in the 19th century, between 1840 and 1890, water power was replaced throughout industry by steam power, which was more versatile and allowed industry more locational options. The steam engine also powered the railroad and steamship expansion, a revolution in transportation (Temin, 1966). Older manufacturing centers such as Lowell and Waterbury declined and others located near ports and railheads took their places. Then, between 1880 and 1930, steam power lost out to electric power and another new industrial regime came into being. Electricity made industry more productive, not only in energy use but also in the use of capital and labor, and stimulated economic growth (Devine, 1983).

At the end of the 19th century, before motor vehicles replaced animal traction, there were almost 14 million horses, mostly draft animals, in America and a million horses in New York City alone. Almost all of them are gone, but we still measure an engine's capacity in horsepower. Between 1900, when production of gasoline-powered autos began in the US, and the 1940s, automotive transportation replaced animal traction. It created a new industry, the petroleum industry, and greatly increased demand for steel, rubber and other products. It stimulated investment in highways and residential construction. It supplanted not only animal traction but also inter-urban rail and urban trolley systems. Gasoline-powered trucks had a similar impact on transportation of goods. The new trucks were three times faster than the horse-powered drays, took less room to store and eliminated the problem of manure disposal (Kay, 1997).

In each transition, there were realignments and disruptions, a process the great economist Josef Schumpeter called 'creative destruction'. What happened to all those farriers and liverymen? Whalers? Steamfitters? In each transition, the industries under threat raised alarms and fought rear-guard battles. In the early days of motor vehicles, some localities required automobiles to be preceded by signalmen waving warning flags. Opponents of railways warned of fires set by engine sparks; opponents of stream power warned of boiler explosions. But, in each case, the energy transition gave rise to the growth of gigantic new industries, new waves of investment and innovation, and an upsurge in prosperity. The economic benefits greatly exceeded the costs. The next surge will be powered by clean energy (von Weizsäcker et al, 2009).

America can meet the climate challenge with technologies already available and in use, and feel confident that a good climate policy stimulus will result in streams of innovation and technological progress. By raising energy efficiency in our houses, offices, appliances and vehicles, and making the transition to abundant domestic non-fossil energy sources, carbon emissions can be reduced by 80 per cent by mid-century. By taking vigorous action to reduce greenhouse gas emissions, the US can persuade other countries to do the same and the potential catastrophe of global warming can be averted.

In doing so, oil imports can be eliminated, even though the US now consumes 25 per cent of the world total and now has only 3 per cent of the world's reserves, numbers that expose the mistaken view that we can drill our way to energy independence. Instead, the economy can rely on abundant renewable energy sources that will become less expensive over time. Solar and wind power costs have both fallen by 90 per cent since 1970 and are expected to fall by another 50 per cent in the next 20 years because of technological improvements and the economies of large scale production. Similarly, the costs of ethanol and biodiesel derived from plants are expected to drop by 50 per cent in the next 10 years.

The US can drastically decentralize energy production. A large fraction of future energy can be produced on rooftops, in parking lots and in farmers' fields all over the country. Energy production can be made invulnerable to large scale shocks or disruptions. It will also be cleaner. Air quality can be improved substantially by reducing emissions of mercury, sulfur and other pollutants from coal-fired power plants, saving tens of thousands of lives every year. The environmental impacts of coal mining and oil spills can be reduced.

The efficiency with which energy is used can be dramatically increased while saving money for households, industries and businesses. Investments in improved energy efficiency can provide an attractive rate of return and also important new employment and business opportunities. Experience in other industrialized countries supports these conclusions. Japan prospered while using only half as much energy per dollar of GDP as the US. Denmark prospers while depending on wind and other renewable resources for 20 per cent of its energy consumption. Western Europe prospers despite prices for liquid fuels that are twice as high as those in America. Studies show that low energy prices are associated neither with higher per capita incomes nor with faster rates of economic growth. They are associated only with low levels of energy efficiency (Kosmo, 1987).

THE ENERGY TRANSITION

This response requires a shift away from fossil fuels toward renewable domestic energy sources and a marked increase in energy efficiency (National Research Council, 2008). Not only is this possible, it is already well under way and needs only to be supported and accelerated. Improved energy efficiency has been the largest source of energy over the past 25 years. Absent those improvements, the US would have had to construct many more power plants and would have burned much more coal, oil and gas than it actually did. Renewable energy industries have been growing very rapidly, at rates of 20 per cent per year or more, many times more rapidly than the mature fossil fuel industries. They still contribute only a small percentage of the total power supply and it will be decades before they become dominant, but that's all right, the transition can and will take decades.

Private and public sector actors throughout the country are already deeply engaged in implementing the transition. State and local governments have enacted standards and other measures to increase the contribution of renewable energy and have developed programs to raise energy efficiency in buildings and equipment. The State of California, through innovative and determined programs, has demonstrated the feasibility of growing in population and economic production over decades while keeping energy use constant. Businesses ranging from the largest multinational corporations to the smallest start-ups are already improving their energy efficiency and marketing an awesome range of products and technologies that will help achieve the transition. Financial markets are pouring investment capital into so-called 'cleantech' ventures and devising financial products, such as location-efficient mortgages and energy-efficient mortgages, which take account of borrowers' improved ability to pay when they buy energy-efficient houses near their places of work. On almost a daily basis, new possibilities are emerging from research labs and universities.

Because of these efforts, the US is poised for a rapid acceleration of the transition. As the costs of new energy technologies are brought down and the costs of the carbon-intensive technologies rise in the marketplace, a tipping point will be reached. From that point on, there will be an even more rapid diffusion and penetration of renewable energy systems.

In decades to come, most electricity will be generated from wind, solar, wave and other clean renewable domestic sources. These industries are projected to triple in revenues just in this decade (Makower et al, 2009). Both inland and offshore wind energy will continue to expand rapidly, aided by new developments in transmission, distribution and energy storage. The potential wind energy resources in the Midwest alone, from Texas to the Dakotas, are several times the nation's current electricity generation. Farmers and ranchers on marginal cropland will benefit from payments for wind towers on their properties and will grow energy crops such as switchgrass – one of the native grasses of the Great Plains – beneath them. New transmission lines will bring wind electricity to load centers and integrate it with other generating sources. Denmark, where wind already accounts for 20 per cent of total electricity, has found that the costs of integrating even such large intermittent wind resources into the grid are small, from a quarter to a half cent per kilowatt hour, partly because as wind farms expand geographically, their fluctuations in electricity production average out. Experts have found that in the US, even with the current transmission and distribution system, 'up to 20 per cent penetration of wind power can be accommodated with no significant operating issues' (Milligan et al, 2009). Advances in energy storage technologies, such as pumped storage of water or compressed air, will also reduce problems of intermittent winds. Wind farms offshore will also become common, as they already are in Europe, generating electricity close to cities and dense population clusters along the coasts. Since wind energy is immune to fluctuations in fuel prices, it will contribute to stability in power prices, which can already be as low as 4–6 cents per kilowatt hour at good sites.

The potential solar energy resource is even larger and costs will continue to decline. Since solar energy can often be generated where it is needed, it competes not with wholesale but with retail electricity rates and is already cost-effective for many users. When the real costs of carbon are included in electricity prices, solar energy will become much more widely competitive. For example, more and more businesses are already leasing out the rooftops of their stores and warehouses and are purchasing solar power through long term purchasing agreements from companies that install and manage solar installations and sell electricity, heat and cooling to the building owners. In the US today, there are 2 million acres of flat unshaded rooftops with areas greater than 35,000 square feet on investor-owned buildings that could potentially be used for electricity generation, with a capacity equivalent to almost 300 thousand-megawatt power plants (Inslee and Hendricks, 2008). Solar roofing tiles and shingles, under development for years and now coming onto the market, will be increasingly integrated into new house construction or roof replacement, allowing households both to reduce their electricity purchases and to sell peak power back to the grid. A Norwegian company, working with university scientists, has announced that within five years it will commercialize a thin solar film that can be sprayed or painted onto windows, walls and roofs.

Solar thermal plants totaling two gigawatts of capacity are already under construction in the Southwest, where the sun shines intensely most of the year. Mirrors focus the sun's rays to heat a fluid, which then runs a steam turbine. Solar thermal plants in the desert can provide baseload power, especially if they are configured to store heat for later use or paired with auxiliary gas turbines. These plants complement wind power, since the wind blows more strongly at night when the sun is down, and can help make renewable energy less intermittent. The country's entire electricity load could be generated by such plants in an area the size of a single county in Arizona.

Decades from now, a new generation of safer nuclear plants will be providing baseload power if a repository for high level nuclear waste is finally put into operation, but other sources will also be available. Geothermal plants harvesting heat from deep underground are likely to proliferate, in part because they can be sited through much of the country and will be competitive with coal after carbon costs are priced into the market. Like wind, geothermal generating costs will be stable over time because fuel costs are nil. Further into the future, the enormous energy potential of ocean tides and waves will also be harnessed, using technologies like those already in operation off the coasts of Northern Ireland and Scotland.

There may also be a future for coal as a source of baseload power, though coal is not as cheap as often claimed if the environmental, health and safety costs imposed on society are taken into account (National Research Council, 2009). The old, dirty and inefficient coal-fired thermal plants still in operation today will be retired because they will be uncompetitive when carbon emissions are priced into the marketplace.

But new, more efficient coal plants using gasification and combined-cycle technology and able to separate and capture carbon dioxide exhaust gases may take their place if it proves possible to sequester large volumes of captured carbon dioxide permanently under the ground or in the deep ocean. Limited volumes of CO_2 are already captured and injected underground to boost production in old oilfields, and power plant exhausts are already being sequestered in this way in Norway. Geological research and testing are well under way. There is enough storage capacity in depleted oil and gas wells in reservoirs that held up for millennia to accommodate the carbon dioxide that could be emitted for 40 years by today's coal-fired power plants. Even more exciting possibilities are under development: one company is making 'green' cement by bubbling carbon dioxide through sea water.

For peaking power, hydroelectric and efficient gas plants will still have a role, but peak power requirements will be transformed by innovations in energy management. The so-called 'smart grid', which embeds information technology into electricity transmission and distribution networks, will support 'time of day' electricity pricing and allow consumers not only to shift their electricity use to off-peak hours but also to sell electricity back to the utility from car batteries and solar installations when its price is high. Many large industrial users already have this energy management capability and utilities are starting to build out the smart grid in residential communities. Under proper incentives, investment in upgrading the distribution network will spread these capabilities throughout the country.

The enormous energy wastage in America's electricity sector will be greatly reduced. Amazingly, two-thirds of the energy used to generate electricity is simply dissipated in cooling towers and cooling waters as waste heat. In other countries, this energy is put to use for heating and cooling. A thousand cities around the world pipe heated water from power plants to buildings, but only a few American cities do so. Combined heat and power plants are much more commonly found today in hospital complexes, universities, commercial establishments and industrial facilities than in electric utilities. The unrealized potential to make use of this almost free energy is huge (Ayres and Ayres, 2009). The reason is lack of incentive, not lack of opportunity. In the future this enormous wasted resource will be productively used.

Over the coming decades, transportation will become much more energy-efficient and largely independent of petroleum products (Cambridge Systematics, 2009). The rise in gasoline prices is already causing a dramatic shift in the automobile marketplace. Gas-guzzling trucks and SUVs have become less popular and consumers are buying more fuel-efficient vehicles, including those with advanced 4-cylinder engines. In years to come much more energy-efficient cars, including new clean diesels, will appear in the US market, paralleling the experience of Europe and Japan. Hybrid electric vehicles are already becoming popular because of their fuel efficiency, but the next generation will be plug-in hybrid vehicles with batteries that can be recharged from an ordinary electric socket, using about the power needed to run a dishwasher. GM plans to introduce its electric car in 2011 and is co-operating with the Electric Power Research Institute to ensure that the electricity grid will then be fully able to accommodate them. Toyota plans to introduce its plug-in hybrid in 2011. With a battery range of 40 miles, much farther than most automotive trips, these cars will run mostly on electricity, lowering driving costs by more than 50 per cent and substantially reducing carbon emissions. Batteries will usually be recharged at night using cheap off-peak power and that power will be generated mostly by wind energy, making use of night-time winds that would otherwise be wasted.

In addition, the 5 million acres of parking lots surrounding our shopping malls and office parks will shade cars under solar panels, the way Google's office parking lot now does. This will not only allow commuters to recharge their vehicles while they work or shop, but will also allow them to use the coming 'smart grid' to sell surplus power back to the utility during peak daylight hours. Instead of paying to park, people will be paid to park. The electricity storage capacity of America's vehicle fleet is several times larger than our entire generating capacity, and, on average, cars are parked 90 per cent of the time, allowing their batteries to be used to improve the reliability and reduce the peak capacity requirements of the electricity system.

When plug-in hybrids do need to run on the internal combustion engine, they will use a fuel consisting of 85 per cent ethanol as most vehicles in Brazil do now. Many cars on the road in the US today can run on E85, often unbeknown to their owners, and the additional costs of modifying engines to do so is only about $100. Ethanol will not be produced inefficiently from corn or other food crops, as it is today, but from switchgrass, agricultural wastes and from other even more productive sources of biomass. Water hyacinth, which produces more biomass in two weeks than switchgrass does in a year, will be grown in municipal and industrial wastewaters and will very efficiently clean them of organic and inorganic pollutants. Algal farms, which require only sunlight, water – even saline water – and carbon dioxide to produce even more biomass per acre, will be grown on the carbon dioxide captured from industrial and utility furnaces (Pienkos, 2007). Such farms sited on unusable land covering only 0.2 per cent of the total land area could produce enough ethanol to run all the vehicles in America. Missouri has recently invested more than $100 million to construct an algal bioenergy plant.

There will be broader transformations of the transportation system. Along densely populated corridors such as the Northeast Seaboard, investments in railways will finally provide rapid and comfortable service comparable to what is already available in Europe and Japan, and is under construction in China. Cities will invest in public transportation, especially light rail and bus service, encouraging people to leave the car at home. Cities will see the wisdom of providing bike paths, bike lanes and bike parking areas, enabling city dwellers to get around town quickly and safely. As a result, American cities will become more vital and desirable places in which to live, contributing to a much needed urban renaissance.

In future years, the economy will have exploited much more of its enormous reservoir of high-payoff-potential energy efficiency improvements (Earhardt-Martinez and Laitner, 2008). For example, compact fluorescent bulbs have so far captured only five per cent of the lighting market, despite their large lifetime savings and better energy efficiency, and even more efficient LED lighting is just beginning to make inroads. Energy-Star rated household appliances have captured only one-third of the potential market; heating and air-conditioning equipment also only one-third; doors and windows less than one-half. The average personal computer today uses three times as much energy as the most energy-efficient models. Standby power devices in television sets, computers and other devices typically use 25 times more power than necessary. The Electric Power Research Institute estimated that efficiency gains could reduce the growth in power use by up to 36 per cent between now and 2030 (Electric Power Research Institute, 2009). Buildings themselves are far less energy-efficient than those in Northern Europe and use three to four times the energy of those being built today under LEED standards, largely because of inadequate insulation, poor site orientation, design and construction, and inefficient lighting, heating and ventilation systems. Industries have not nearly exploited all their energy-saving opportunities (Shipley and Elliott, 2006). For example, the US uses 50 per cent more energy to produce a ton of cement than Japan does and twice as much energy to produce a ton of paper.

Over time this large efficiency gap will be reduced, raising economic productivity and reducing energy demands even as technological progress makes further improvements possible. These energy efficiency gains are 'the low-hanging fruit', but fortunately the low-hanging fruit, once harvested, grows back with further technological progress. Even with today's technologies, a substantial fraction of emissions could be eliminated through cost-saving efficiency improvements (McKinsey & Co, 2007). Innovations continue to emerge, such as a new cement that, once installed, scavenges CO_2 from the atmosphere and a zero-carbon sheetrock that can be manufactured without using high heat.

ACHIEVING THE TRANSITION

This transition can be achieved with technologies already in use or in sight, never mind the innovations sure to come (Pacala and Socolow, 2004). It requires only institutional change and political commitment – in a word, governance. That's the bad news: America is much better in technology than in governance.

The first essential requirement is to undo the legacy of the past, which has been shaped by abundant fossil energy, low energy prices and few environmental constraints. Those conditions no longer exist but their consequences linger on. Our energy system is outmoded and so are the government policies that have supported it. The fossil energy industries continue to enjoy by far the largest share of energy-related subsidies and tax concessions. Government support for research, development and deployment also goes overwhelmingly to the technologies of yesterday, not to those of the future. The regulatory structure is still not designed to support and accelerate the energy transition that is urgently needed. Energy policy needs to be redirected as urgently as do other aspects of the energy system.

The first of these legacies from the past is an economy of stunning energy inefficiency. It has been calculated that only one per cent of the energy generated by the average car is actually used to move the driver, who typically travels in splendid isolation. In addition to the conversion, friction and other losses, most effective energy is used to move the enormous weight of the vehicle, usually 25 times that of the person behind the wheel.

When the average light bulb is switched on, only 3 per cent of the energy used provides light. 90 per cent of the electricity used by an incandescent bulb generates heat, not light, and only 30 per cent of the energy used to generate and transmit the electricity powers the bulb. Our appliances and electronic devices use approximately a trillion kilowatt hours of electricity per year, about the output of seven 1000-megawatt power plants – while they are turned off. Most American houses lack proper insulation and leak energy like sieves. These and other glaring inefficiencies are legacies of a profligate energy past and can and must be corrected (Gellings et al, 2006).

The most appalling waste is the two-thirds of energy in the fuels used to generate electricity that is dissipated in cooling towers and waters. The average electricity generating plant's efficiency hasn't improved in almost a half century. Even China's electricity sector is more efficient than ours. Denmark's performance is almost twice as good, mainly because much of the low grade energy in power plant exhausts is used for heating purposes. The energy that our electric power industry discards each year as waste heat is more than the entire Japanese economy uses.

How could that have happened and been allowed to continue? Well, for starters, in the two-thirds of States in which the power sector is still subject to rate of return regulations, regulations have provided almost no incentives for utilities to conserve fuel. Utilities are allowed almost automatically to pass on their fuel costs to customers, recouping all fuel expenditures. If your fuel bills were automatically reimbursed, how deeply would you care about your house's energy efficiency? State public utilities commissions should link fuel cost recovery to the generating plant's heat rate (BTUs burned per kWh generated): the greater the plant's efficiency, the greater the fuel cost pass-through. In addition, why not allow the electric utilities unlimited profits on any of the low energy heat it can use or sell? Though the electricity sector was thought to be a natural monopoly, electricity competes with gas, fuel oil and other technologies in markets for heat and cooling.

In those States still subject to regulation, electric utilities still enjoy an effective monopoly on sales, based on the obsolete theory that economies of scale make them natural monopolies. Now, smaller and cleaner gas-fired combined cycle plants are competitive for peaking purposes and can be located close to cities and industrial areas that might make use of their low grade heat more easily. Cogeneration plants that sell both heat and electricity are potentially competitive but are forbidden by law to sell electricity to anyone but the electric utility, even to a potential customer right across the street (Casten and Collins, 2005). PURPA, the Public Utilities Regulatory Policy Act passed in 1978, tried to remedy some of this by mandating that utilities offer to buy electricity from such independent producers at their 'avoided cost'. This has provided some stimulus, but utilities have little incentive to comply, since they are guaranteed a rate of return on their own investments, and so have dragged their feet by using estimates of marginal costs, transmission costs, inter-connection charges and back-up charges to lower their calculated avoided costs. It would be more effective for PUCs to stipulate 'feed-in tariffs' for carbon-efficient independent power producers at reasonable levels, as several European nations do.

Another barrier was built into the Clean Air Act, passed in 1970, which exempted pre-existing power plants from stricter pollution control requirements, on the assumption that they would soon be retired. Recognizing that they had created a cost incentive to keep these plants going, in 1977 Congress amended the Act to require plants undergoing modernization or rehabilitation investments to adopt standards applicable to new plants. Routine maintenance investments were exempted from this requirement, however. Utilities responded, if at all, by classifying the power plant equivalent of kidney transplants to keep the patient alive as 'routine maintenance'. Though Clinton's EPA brought some of them to court, the Bush administration subsequently gutted the regulations and dropped the lawsuits. Consequently, some of the nation's least efficient and dirtiest plants, built 40 years ago or more, are still wasting energy and spewing pollutants into the air. Resuscitating and enforcing new source review requirements would lead utilities to retire many of these clunkers rather than spend the money to improve them. China has been shutting down and replacing its least efficient power plants. Why can't we?

Utilities have also had little incentive to promote 'demand side' energy efficiency among customers, even though they are in the best position to do so, because the more energy they sell, the more money they make, at least between intermittent rate hearings. Traditionally, rate of return regulations imposed by State public utility commissions have allowed utilities a mark-up over costs to provide an allowable return on investment. Average cost pricing, which fails to reflect the much higher costs of electricity in periods of peak demand, has also blunted customers' interest in energy efficiency. Fortunately, this situation has been changing. Many public utility commissions have mandated that utilities carry out demand side management programs and have allowed a reasonable rate of return on those program costs, or have allowed independent energy service companies to manage those programs. Some States have imposed energy efficiency standards, requiring utilities to meet improvement goals. Most utilities have responded with limited enthusiasm, even though studies show that end-use efficiency gains come at about half the cost of new generation (Hoppock et al, 2008).

Today, at least a dozen States are going further by 'decoupling' utility revenues from the volume of sales so that utilities become equally happy serving customers by helping them reduce demands or by increasing supply. The simplest way to do this is to grant utilities a specified amount of revenue per customer, regardless of the amount of electricity sold, and adjusting rates accordingly to offset deviations of actual from expected sales. This provides utilities with the incentive to promote efficiency gains if doing so is cheaper than supplying more electricity, as it usually is. Utility regulation nationwide should be revised along such lines to provide electricity suppliers strong incentives to pursue efficiency.

Moreover, utilities have been finding that information technology incorporated into transmission and distributions systems, which are still essentially like the one designed by Thomas Edison nearly a century ago, pays for itself just in utility cost reductions. It also facilitates 'time of day' electricity pricing and allows consumers much greater control over their demand patterns. Decoupling, the smart grid and similar innovations should be vigorously promoted nationwide. In 2007 Congress gave NIST responsibility to develop consensus interoperability standards for the smart grid by December 2008. By May 2009, however, NIST had only completed the first of three phases of this work and the Energy Department had just announced an expanded grant program for smart grid development.

Similarly, all public utility commissions should require utilities to carry out integrated resource planning before proposing to build new power plants. Such planning obligates the utilities to consider potential efficiency improvements as an alternative to increasing supply. More often than not, efficiency is the cheaper option and will save ratepayers money while accelerating the transition. Resource planning should also require that utilities assign a reasonable price to future carbon emissions from proposed new plants. If they don't, they might build additional coal-fired plants that will soon be uneconomic when climate policies are enacted, leaving ratepayers with stranded costs (US Environmental Protection Agency, 2008).

Every plausible scenario of the energy transition includes a role for a new generation of safer nuclear plants, like those that are now being built in other countries. One obstacle such investments must overcome in the US is that of storing the depleted nuclear fuel, which remains highly radioactive. Today, more than 60,000 tons of such material is in 'temporary' storage at 104 plant sites around the country, and the total amount in storage is rising by about 2000 tons per year. More than 160 million people live within 75 miles of these plants. The site designated for a permanent repository in Yucca Mountain, in a remote and almost unpopulated area of Nevada, was apparently abandoned by the Obama administration at the behest of the Senate majority leader, Harry Reid, who faced a tough re-election battle.[1] Nevadans raised concerns about the safety of radioactive wastes being transported to the site. What about the safety of wastes sitting in pools of water all over the country?

Yucca Mountain is the mother of all NIMBYs. If a repository cannot be sited there, it is difficult to think where one might be put. The technical criteria for the site include assurance that the repository would be secure for 10,000 years, particularly from water seepage that might corrode the special steel containment vaults. Yucca Mountain is about 75 miles from Death Valley, probably the driest spot on the continent. This exemplifies the incoherence of current energy policy. America faces a national and global ecological disaster that could spin out of control within this century and an important part of the solution is hostage to local concerns about remote risks that might materialize over ten millennia.

Similar governance problems impede the more rapid development of renewable energy. The abundant wind resources of the Great Plains must be brought to load centers through new high voltage transmission lines. Siting and financing such interstate transmission lines have been held up for years by controversies among federal agencies, State governments and public utility commissions, electric utilities, renewable energy developers and consumer groups regarding who should pay for them, and by meta-controversies over who should decide who should pay. The Constitution grants the Federal Government power to regulate interstate commerce, squarely to resolve controversies like this one, and in 2005 the Energy Policy Act authorized the Federal Energy Regulatory Commission (FERC) to use powers of eminent domain to site transmission lines within 'National Interest Electric Transmission Corridors', if necessary. The FERC should recognize the necessity of bringing abundant renewable energy to market and should make decisions to resolve these controversies. The analogous Interstate Highway System was created by the Federal Government in the Eisenhower administration and financed largely by taxes on vehicle fuels. Why shouldn't an interstate transmission system be financed by federal electricity taxes? The benefits will accrue to all consumers, not only through cleaner and ultimately cheaper electricity but also through better air quality and reduced risks from climate damages.

Offshore wind development has similarly been impeded by lack of coordination and controversies within the Federal Government between the FERC, the Minerals Management Service, the Coast Guard and other agencies, and with State and local governments and their representatives. The first proposed development, Cape Wind, was held up for nine years, largely because Senator Edward Kennedy disliked the prospect of seeing dime-sized wind towers on the horizon from his back deck in Hyannisport, and by similar sentiments among the wealthy and influential summer homeowners on Nantucket and the Vineyard. These folks revived a historical alliance with the Wampanoags, dormant since King Phillip's War, inciting them to claim that their cultural rituals required viewing the unimpaired sunrise across Nantucket Sound.[2] Only in 2010 did the Secretary of the Interior approve the project, by which time more than 800 offshore wind towers were already in operation in Europe. For this transition to proceed, there must be a greater sense of national priority and urgency.

For example, federal energy subsidies in general are strongly in need of reform. Between 1999 and 2007, such subsidies doubled and now cost more than $16 billion per year. They encourage energy consumption or production or both, but only 35 per cent of that total amount supports renewable energy or energy conservation. Ironically, subsidies for alternative energy have been hijacked by the coal industry for production of 'refined' coal, lightly treated with chemicals. Subsidies for refined coal are twice those for all truly renewable energy alternatives combined. Subsidies should be concentrated on energy efficiency and the technologies of the transition, including advanced transmission and distribution systems as well as renewables. The stimulus package funded in the 2009 American Recovery and Investment Act took a step in that direction (ICF International, 2009), but less than 10 per cent of that funding went to 'green' energy projects. In Europe, the proportion of stimulus spending going in that direction was 38 per cent; in South Korea, it was more than 80 per cent (Robins and Clover, 2009).

America's transportation and land use system is predicated on the automobile and cheap gasoline. For decades automakers devoted all their technological innovations to the goal of increasing the power and weight of their vehicles, keeping fuel efficiency unchanged. As a result, in 2006 the average miles per gallon achieved by Ford's fleet of vehicles was less than that achieved by Ford's Model T 80 years earlier. Detroit automakers' bet on SUVs and pickup trucks instead of hybrids became ugly when fuel prices rose and contributed to their bankruptcies. Still, across America, commuters drive to work from bedroom suburbs, usually alone, suffering pollution and increasing highway congestion, while inner cities deteriorate and public transportation options stagnate. Congress has recently raised fuel efficiency standards, but they are still weaker than those enacted in China. A carbon efficiency standard for automotive fuels would be a more powerful stimulus to biofuels and other alternatives but would be tricky to assess on a life cycle basis (Peña, 2008).

In the transportation area, government support also needs to be shifted more strongly toward mass transit. Recent federal legislation has moved in this direction and has provided more flexibility to State and local governments in allocating available funds, but still the Mass Transit Account receives only about 15 per cent of funds generated by transportation taxes. By far most of the rest supports highway construction. For decades, well after the Interstate Highway System was completed, gasoline taxes have funded highway construction, which has encouraged more driving, which has generated more tax revenues for more highway construction. Obviously, a transition demands a major reorientation of transportation funding toward urban mass transit and bicycle travel, and away from new highway construction. In European cities such as Vienna, Amsterdam and Copenhagen, a large percentage of trips are by bicycle. Bicycle lanes require only parking bans on one side of designated streets and lines painted on pavements. How hard is that? A good way to finance these changes would be through congestion pricing, which has worked well in London, Singapore and elsewhere. Vehicles entering the central city during rush hours pay tolls (or higher tolls). This combats worsening congestion, which by 2007 already cost the nation $78 billion per year in lost time and 2.9 billion gallons of wasted fuel. Mayor Bloomberg tried such a program in New York City, which already has tolls on bridges and tunnels into Manhattan, but local politics stymied the effort.

Along these lines, the US could adopt a 'fee-bate' system for purchases of cars and other energy-intensive products, as China has done: excise taxes are levied on gas guzzlers and the revenues are used for tax credits on purchases of energy-efficient vehicles. The financial constraint that leads many households to purchase inefficient appliances and forgo energy upgrades with short payback periods because of the initial cost hurdle can be overcome by loan programs that spread the cost over time so that they can be offset by savings in energy expenditure. Innovative measures such as these should be adopted nationally as part of the effort to achieve the needed energy transition as quickly as possible.

The push for alternative fuels also became just another boondoggle for the farm States that are so grossly over-represented in the Senate. Corn ethanol barely contributes any net reduction in greenhouse gas emissions, but Congress has not only encouraged it but also levied a $0.51 per gallon tariff on much more cost-effective sugarcane ethanol imports from Brazil and elsewhere. All ethanol mandates and subsidies should be shifted from food crops, such as corn, soy and palm oil, toward agricultural residues, switchgrass, algae and other superior sources.

Governments can and are doing even much more than this to promote a rapid market transition. Studies have shown that consumers do not always make their most financially advantageous spending decisions or follow through on their espoused environmental preferences. The reasons include inattention, inertia, lack of information and real financial constraints. Consumers often need a 'nudge' to act in the right direction. For example, Walmart greatly increased the sale of compact fluorescent bulbs just by displaying them more prominently than incandescent bulbs. Governments can make the more energy-efficient and cost-effective decision the default choice by setting minimum efficiency standards. The Energy Policy Act of 2005 and the Energy Independence and Security Act of 2007, along with numerous State and local enactments, have gone a good distance in this direction. The Energy Department is supposed to roll out dozens of revised appliance standards over the next few years and should do so promptly, despite industry lobbying. Studies show that such standards will save both money and energy (Meyers et al, 2008).

The next step would be to improve the woeful inefficiency of buildings, which consume one-third of all energy. Energy use in many buildings is affected by a classic failure of economic incentives: tenants and owners usually have little say over their buildings' design and construction and can't see how well structures are sealed and insulated after the siding and sheetrock are installed, while architects, developers and contractors don't pay the occupants' utility bills. Moreover, residential developers make higher profit margins by selling bigger houses, since they are really selling enclosed space. Often, the result is stylish but leaky houses with 'great rooms' and cathedral ceilings that require excessive energy to heat and cool. In cities across the country there are still hundreds of thousands of apartment buildings with only one central meter, which forces owners to incorporate utility costs into the rent and thereby encourages tenants to overuse heat and air conditioning. Buildings in Northern Europe and Japan use far less energy per square foot than ours do. The answer to such incentive failures is disclosure and regulation. The Waxman–Markey bill moved in this direction by proposing a national building code embodying energy efficiency provisions – a good idea, but it has not yet become law.

In addition to removing these perverse incentives, government action can positively encourage market development and expansion in support of the energy transition. Government procurement is a good example. Accounting for 20 per cent of gross domestic product, government spending has a significant direct effect on US carbon emissions. Shockingly, a large percentage of public spending decisions are not based on life cycle cost analysis but on the low initial cost bid or some other criterion. As a result, the government procures energy-inefficient buildings and equipment that waste energy and taxpayers' money. Life-cycle cost analysis that includes a carbon price should be mandatory for all significant government procurement decisions and building acquisitions. Studies have shown that buildings designed to LEED standards are not only cheaper but save even more money by raising the productivity of employees working in them. Thus government action could save energy and money directly and also build markets and help achieve economies of scale for suppliers of energy-efficient products.

Federal support for energy research, development and deployment must also increase and be redirected toward the energy technologies of the future. In inflation-corrected dollar terms, federal energy research is less than one-quarter what it was 30 years ago, though the need today is greater. Research on the smart grid system, on batteries, superconductors and other energy storage devices, is of high priority. Research on the safety and capacity of repositories for captured and sequestered carbon dioxide is crucial. Research on advanced energy conversion devices such as fuel cells and low cost solar photovoltaics is also important. Obviously, such research should be carried out in partnership with private industry and reward commercialization and deployment rather than just R&D spending. For example, MIT scientists have recently discovered a new catalytic process essential to fuel cells that doesn't need expensive platinum and can operate in ordinary ambient conditions. The payoff is enormous if discoveries like these can quickly be developed to the commercial stage and brought to the market.

The second essential requirement is to enlist and mobilize the powers of the marketplace. Businesses are devising and marketing an impressive array of new products and services designed to improve the energy performance of their customers or to make alternative energy technologies more competitive. Companies large and small are already busy improving the energy efficiency of their own operations and facilities and making attractive returns on these largely risk-free investments. Typically, though, energy saving is not considered part of the core business, and energy-saving investments must promise a two- or three-year payback, implying internal rates of return exceeding 35 per cent, despite their low risk. That is one reason why so many opportunities remain to be harvested.

All this activity and innovation are taking place even before policies are in place that correct for the real economic costs of greenhouse gas emissions, just in anticipation of a future market price or penalty for carbon emissions. Once effective policies are enacted that create an enduring market price for carbon emissions, all businesses will be motivated to calculate the potential profitability of innovations that reduce use of fossil fuel, and new products that contribute to reducing carbon emissions. The ingenuity and multiple talents and resources of American industry will be mobilized in search of these new sources of profits, giving rise to a wave of investment and innovation the scope and magnitude of which is impossible to foresee.

Consumers will also respond to the changes in energy prices and options. Pessimists often claim that consumers are sluggish and unresponsive to price signals, but the facts demonstrate otherwise. The increases in fuel prices in recent years have given rise to large, rapid changes in consumer behavior: a sharp drop in the purchase of gas-guzzling vehicles, changes in driving habits, changes in the use of public transport, changes in tele-commuting, even preferences for urban over suburban housing. Consumers do respond to market signals, especially when accompanied by supporting marketing and market-facilitating measures.

Government policy must create those market signals. Climate change and the security risks of energy dependence are massive market failures, in that their costs are not borne directly by those who contribute to the problem. It must be the role of government to correct this market failure by enacting a price for carbon emissions, a surcharge on the use of fossil fuels that reflects these very real economic costs. How best to do this is explained in the following chapter.

CONCLUSION

In order to emphasize the need for a vigorous national effort to meet the energy challenge, analogies have been drawn to the Apollo Project, the race to put a man on the Moon, and to the Manhattan Project, which led to the creation of the atomic bomb. The truth is, though, that this challenge is much broader and far more urgent. The Manhattan Project involved a small group of scientists operating in secret and required some specialized manufacturing capabilities. Had it not succeeded on schedule, World War II would still undoubtedly have been won. The Apollo Project was also limited in scope and its success was only a matter of national prestige. The energy challenge demands that the entire nation and all its economic capabilities be engaged: households, businesses large and small, and all levels and branches of government. The consequence of failure would be an irreversible ecological and human catastrophe of global dimensions.

Failure is an option but by no means a necessity. The technological and economic abilities to succeed are available, but there is no time to spare. Every year's delay in taking the sorts of actions described above will result in higher atmospheric greenhouse gas concentrations and make it more difficult to stabilize concentrations in the future at a level that could be described as safe. America must face up to the energy challenge and do it now.

NOTES

1 Each of Nevada's two Senators have one constituent for every fourteen represented by a Senator from California.

2 The criteria for membership in the Mashpee Wampanog Tribe read more like criteria for election to the Mashpee Country Club or Chamber of Commerce: 'The following persons are eligible for membership in the Mashpee Wampanoag Tribe:

A. Persons who trace direct lineal descent to:
   (1) a Mashpee Indian identified in the Report to the Governor and Council, concerning the Indians of the Commonwealth, under the Act of 16 April 1859 written by John Milton and published in 1861 by William White, Printer to the State, in Boston, Massachusetts (hereinafter 'Earle Report'); or
   (2) the 19th century unions of Georgina Palmer and Charles Peters or Leander Peters and Lydia DeGrasse;
B. Persons who demonstrate tribal community involvement;
C. Persons who have not publicly denounced Mashpee tribal existence or their affiliation to the Tribe; and
D. Persons who have lived in or near Mashpee, Massachusetts, or have had family members actively involved in tribal community affairs who have lived in or near Mashpee, Massachusetts, for at least the preceding 20 years prior to application for membership.'

REFERENCES

Ayres, R. and Ayres, E. (2009) *Crossing the Energy Divide*, Wharton School Publishing, NJ

Cambridge Systematics (2009) *Moving Cooler: Transport Strategies to Reduce Greenhouse Gas Emissions*, Urban Land Institute, Cambridge, MA

Casten, T. and Collins, M. (2005) *Optimizing Future Heat and Power Generation*, World Alliance for Distributed Energy, www.localpower.org

Devine, W. (1983) 'From shafts to wires: Historical perspective on electrification', *Journal of Economic History*, vol 43, pp347–362

Earhardt-Martinez, K. and Laitner, J. (2008) 'The size of the US energy efficiency market', Report E083, American Council for an Energy Efficient Economy, www.aceee.org/research-report/e083

Electric Power Research Institute (2009) *Assessment of Achievable Potential from Energy Efficiency Demand Response Programs in the US (2010–2030)*, Palo Alto, CA

Gellings, C., Wikler, G. and Ghosh, D. (2006) 'Assessment of US electric end-use efficiency potential', *Electricity Journal*, vol 19, pp55–69

Grubler, A. (2008) 'Energy transitions', *Encyclopedia of Earth*, last updated 3 June 2008, www.eoearth.org/article/Energy_transitions

Hoppock, D., Monast, J. and Williams, E. (2008) *Transforming Utility and Rate-payer Support for Electrical Energy Efficiency Nationwide*, Nicholas School for the Environment, University of North Carolina, Chapel Hill, NC

ICF International (2009) 'Greenhouse Gas Implications of the Stimulus Package', Greenpeace, Washington, DC

Inslee, J. and Hendricks, B. (2008) *Apollo's Fire: Igniting America's Clean Energy Economy*, Island Press, Washington, DC

Kay, J. (1997) *Asphalt Nation*, University of California Press, Berkeley, CA

Kosmo, M. (1987) '*Money to Burn?*', World Resources Institute, Washington, DC

Makower, M., Pernick, R. and Wilder, C. (2009) *Clean Energy Trends 2009*, Clean Edge, Inc., San Francisco, CA

McKinsey & Co (2007) '*Reducing US Greenhouse Gas Emissions: How Much and at What Cost?*', New York

Meyers, S., McMahon, J. and Atkinson, B. (2008) *Realized and Projected Impacts of US Energy Efficiency Standards for Residential and Commercial Appliances*, Lawrence Berkeley Laboratory, Berkeley, CA

Milligan, M., Lew, D., Corbus, D., Piwko, R., Miller, N., Clark, K., Jordan, G., Freeman, L., Zavadil, B. and Schuerger, M. (2009) 'Large-scale wind integration in the US: Preliminary results', *Conference Papers of the National Renewable Energy Lab*, National Renewable Energy Lab, Golden, CO

National Research Council (2008) *America's Energy Future*, National Academy Press, Washington, DC

National Research Council (2009) *Hidden Costs of Energy: Unpriced Consequences of Energy Production and Use*, National Academy Press, Washington, DC

Pacala, S. and Socolow, R. (2004) 'Stabilization wedges: Solving the climate problem for the next 50 years with current technology', *Science*, vol 35, pp968–972

Peña, N. (2008) *Biofuels for Transportation: A Climate Perspective*, Pew Center on Global Climate Change, Washington, DC.

Pienkos, P. (2007) 'The potential for biofuels from algae', National Renewable Energy Lab, Algae Biofuels Summit, San Francisco, CA, 15 November

Robins, N. and Clover, R. (2009) '*A Climate for Recovery: The Color of Stimulus Goes Green*', HSBC, London

Shipley, A. and Elliott, R. (2006) 'Ripe for the picking: Have we exhausted the low-hanging fruit in the industrial sector', American Council for an Energy Efficient Economy, www.aceee.org

Temin, P. (1966), 'Steam and waterpower in the early 19th century', *Journal of Economic History*, vol 26, pp187–205

US Environmental Protection Agency (2008) 'Sector collaborative on energy efficiency accomplishments and next steps', National Action Plan for Energy Efficiency, Washington, DC

von Weizsäcker, E., Hargroves, K., Smith, M., Desha, C. and Stasinopoulos, P. (2009) *Factor Five: Transforming the Global Economy through 80 Per Cent Improvements in Resource Productivity*, Earthscan, London, www.cleanedge.com/reports/reports-trends2009.php

Chapter 3
Choosing the Right Policy Architecture

Let's start by understanding the best framework for a national climate policy, one that can bring about the energy transition described in the last chapter. Then we can examine the effects of compromises and shortfalls from that policy that politics are likely to impose.

The best policy is one that establishes a monetary penalty or cost throughout the economy for emitting greenhouse gases by burning fossil fuels or by other industrial processes (Tietenberg, 2006, Chapter 15). That cost, usually described as a 'price on carbon', must be sufficient to bring about an emissions reduction of about 80 per cent from today's level by 2050. This policy will correct a massive inefficiency in today's marketplace: those who are responsible for emitting greenhouse gases that create the severe costs and risks of climate change do not suffer any commensurate cost or loss themselves and so lack any strong economic incentive for reducing emissions.

A policy that imposes an economy-wide price on greenhouse gas emissions would enlist the best features of the American economy in solving the climate problem: its abilities to innovate, its competitiveness, its dynamism and flexibility. Without governmental regulations, this policy approach would encourage users of fossil fuels to use them more efficiently and to explore alternatives; it would encourage producers of energy-intensive products to find more efficient production processes; it would encourage users of energy-intensive products to buy more efficient models or to find alternatives; it would encourage developers of alternative energy systems and energy-efficient products to bring them to market; and it would offer enticing potential rewards to inventors and entrepreneurs who seek out better products and processes. A policy that puts a national price on greenhouse gas emissions would create an incentive for all businesses and households to help in solving the climate problem in whatever ways are open to them.

This approach would also allow emitters the greatest possible flexibility in deciding what reductions they should make and how and when to make them. If it's cheaper to reduce emissions from one facility than another, companies could do that. If it's cheaper to reduce emissions next year rather than now, companies could do that as well. If it's cheaper to reduce emissions by altering product design or plant operations rather than installing new equipment, companies could do that too. A price on carbon would encourage the most cost-effective ways of reducing emissions.

The absence of such a price on carbon is a serious barrier to investors and developers of alternative technologies, leaving them to cope with uncertainty about the potential returns on any investments they might make. According to a recent inter-agency government report on carbon capture and storage:

> The lack of comprehensive climate change legislation is the key barrier to carbon capture and storage deployment. Without a carbon price and appropriate financial incentives for new technologies, there is no stable framework for investment in low-carbon technologies such as carbon capture and storage. (Inter-agency Task Force on Carbon Capture and Storage, 2010)

The best way to bring about this price on emissions is through an 'upstream' cap and trade system that limits sales of fossil fuels in the US, whether from domestic production or imports. The Federal Government would issue permits to firms that sell fossil fuels. Permits would be required of first sellers of such fuels and would be enforced at the refinery gate in the case of petroleum, the first distribution point in the case of natural gas, at the mine shipping terminus in the case of coal and at the port in the case of imports. Permits would be calibrated to the carbon content of each fossil fuel type and be tradable among first sellers. Reductions in greenhouse gas emissions would be imposed through gradual year-by-year reductions in the permits available.[1] Then, as the supply of fossil fuels in the market gradually diminishes, their prices will rise, establishing a nationwide price on carbon.

Carbon dioxide makes up more than 80 per cent of all US greenhouse gas emissions (US Department of Energy, 2009). This system will be effective in limiting carbon dioxide emissions, since virtually all such emissions arise from the combustion of carbon fuels. By limiting the availability of fossil fuels at their source, all fuel uses will be covered, whether for electric power generation, industry, transportation, or household or commercial energy. Because coverage of carbon fuels will be comprehensive and imposed at source, limitations will be effective and assured. The comprehensive coverage of an upstream cap and trade regime imposed on first sellers of fossil fuels ensures that it will not only be effective but also cost-effective. To maximize the value of their permits, first sellers will give priority to lower cost sources of the fossil fuels they are permitted to sell, and trading will equalize the price per ton of carbon emissions across all fossil fuels.

More importantly, the limitation of supply will drive up fossil fuel prices and the prices of energy-intensive products, providing incentives to all direct and indirect users to reduce fossil fuel use by improving efficiency or reducing low priority uses. Users who can do so at relatively low cost will reduce purchases more; those who cannot will reduce purchases less, so the more essential uses will be preserved and the less essential will be eliminated. This process of economizing on fuel use will take place throughout the economy, ensuring that all low-hanging fruit will be harvested.

Higher prices for fossil fuels will provide a clear economic incentive to develop and deploy alternative energy technologies. Moreover, continuing reductions in the availability of fossil fuels as permit levels are reduced ensures a growing market space for renewable energy and a robust expectation that that market will expand. Investment will flow into alternative energy and costs will decline with increasing scale, research and development, and learning by doing.

Rising energy prices will stimulate energy efficiency technologies and investments throughout the economy, especially in such key sectors as transportation and buildings. In contrast to a tightening of vehicle fuel-economy standards, which perversely encourages car owners to drive their inefficient old cars longer to avoid the higher price of a more efficient new car, an upstream cap and trade system that raises fuel prices will encourage car owners to scrap their inefficient old cars sooner and buy new ones. This approach will be much better for hard-pressed Detroit automakers.

Similarly, energy efficiency innovations will be encouraged in industrial process equipment, motors, industrial controls, heating and cooling equipment, building envelopes, lighting and a host of other energy-using technologies. Investments in 'cleantech', which is already a booming segment of the venture capital market, will attract even more attention as fears of a cleantech 'bubble' are replaced by expectations of long-lasting improvements in investment returns (Woody, 2010).

Because the price on carbon is rooted in mandatory limitations on the first sales of fossil fuels, concerns about the responsiveness of consumers and other energy users to price signals are irrelevant. Though the cuts will drive up energy prices and the prices of energy-intensive goods and services, the price responsiveness or 'demand elasticity' of purchasers will determine only how far prices will rise, not how extensive the reductions will be. Reductions will be predetermined by the availability of permits. Differences in the responses of various energy users will only help in discriminating between more and less essential uses.

Even though an upstream system would be comprehensive in covering all energy uses in the world's largest economy, it would be comparatively simple to monitor and enforce. There will be only about 2000 permit holders whose sales of fossil fuels will be monitored through a paper trail. The Congressional Budget Office, in comparing different policy architectures, concluded that this would be one of the easiest to administer (Congressional Budget Office, 2008). Administrative costs in running an upstream cap and trade regime would be low, both for the government and the private sector, because it would not be necessary to monitor emissions and because fossil fuel sales are already reported and monitored for various other purposes. Sales of domestically produced and imported fuels would be matched annually against permits held by the seller.

This system could be extended to cover large sources of other greenhouse gases, such as methane, by capping emissions from such large emission sources as petroleum refineries, natural gas pipelines and storage facilities where emissions can be monitored accurately. Permits for such emissions would be denominated in carbon dioxide equivalents, taking into account their relative climate-changing potential, which in the case of methane is about 20 times that of carbon dioxide (US Environmental Protection Agency, 2010).

Another advantage of an upstream cap and trade system is its ability to help correct distortions in American energy markets caused by energy subsidies of all kinds: producer subsidies, consumer subsidies and subsidies to competing energy sources, conveyed through all sorts of tax, credit and expenditure vehicles. These subsidies have been established over time, justified as serving many purposes and interests, and persist through lavish lobbying by beneficiaries. Their overall net effect is to raise domestic energy production and consumption. For example, oil extraction at great depths in the Gulf of Mexico, like that behind the Deepwater Horizon disaster, is encouraged by reduced royalty payments on oil extracted at those depths (Wang, 2010). A beauty of an upstream cap and trade regime is that the effects of all energy subsidies on fossil fuel production and consumption are negated. Total production and consumption of fossil fuels would be limited by the overall availability of permits, so energy subsidies would just influence the profits made by various energy sellers and the costs borne by consumers. Though they may also affect the relative market shares of oil, gas and coal in energy markets within the overall carbon limit, subsidies would no longer affect carbon emissions in an upstream cap and trade system.

Moreover, if energy companies are made to pay for their permits in an auction, the Federal Government will be able to recover much of the value of subsidies afforded to energy producers. How much an energy company is willing to pay for a permit to sell another ton of carbon fuels will be determined by the potential profit on that sale. That profit will include the value of the subsidy, so companies getting higher subsidies will bid higher prices for permits, other things being equal. The money the government receives for permit sales will thus include much of the money that it pays out as producer subsidies. At the same time, because sellers of renewable energy don't need to bid for permits, they will retain the value of whatever subsidies they now receive. An upstream cap and trade system offers a way to reform and redirect energy subsidies without attempting a frontal attack on those entrenched interests.

It is ironic that Republicans in Congress and interest groups funded by energy interests are now attempting to demonize this cap and trade approach, as it has been used successfully for decades by both Republican and Democratic administrations in controlling other air pollutants, such as lead and sulfur, and is credited with substantial savings to industry in reducing those emissions (Hahn and Hester, 1989). The administration of George W. Bush proposed expanding the uses of cap and trade approaches to reduce nitrogen oxides and mercury emissions and even for regulating catches in ocean fisheries, because it is effective, efficient and market-friendly. There is already a carbon cap and trade system, the Regional Greenhouse Gas Initiative (Regional Greenhouse Gas Initiative, 2007), in place among the Northeastern State governments and another being put in place on the West Coast, the Western Climate Initiative (Western Climate Initiative, 2009). The US pushed hard and successfully in international climate negotiations for the use of cap and trade mechanisms. The European Union already operates a cap and trade system, which will be intensified after 2012 if the US adopts national constraints. Other Annex I emitters, including Australia, New Zealand and Canada, are moving to install cap and trade systems as well.

FLAWS IN THE MAIN ALTERNATIVES

The other way to create an economy-wide price on carbon would be to initiate a 'carbon tax' by taxing sales of fossil fuels in proportion to their carbon content per BTU. Taxes would be collected on sales of domestic and imported fuels alike. The tax on coal would be highest, about twice as high as the tax on natural gas, reflecting coal's higher carbon content per unit of energy content. The tax would be collected from the roughly 2000 primary sellers of fossil fuels. The tax levied on coal would be partially absorbed by shareholders of coal mining companies, since there are few alternative uses for a coal mining property. To the extent that US demand for petroleum is reduced by the tax, it would also cause a small reduction in world oil prices below their pre-tax levels, since the US is a very significant consumer of petroleum. For the most part, however, the tax would be reflected in higher fuel prices charged to consumers, and in this way would work its way throughout the economy. Taxes on sales of fossil fuels for purposes other than combustion, such as for chemical feedstocks, would be rebated, since those uses don't result in emissions.

A carbon tax would have the same effects on prices of carbon fuels, energy-intensive products and processes as an upstream cap and trade system and would have the same incentive effects throughout the economy. It would also be relatively easy to administer. Yet right-wing Republicans and energy interests have argued that a carbon tax, like all taxes, harms the economy and reduces employment. Politicians wary of openly supporting any tax increase therefore shy away from this alternative (Broder, 2009).

The argument that a carbon tax would harm the economy is complete nonsense, because it doesn't distinguish between a tax that distorts markets and a tax that corrects a market distortion. Failure to charge emitters of greenhouse gases for the damages their actions impose on the economy is a massive market distortion, one that a price on carbon would correct. In principle, taxes should be imposed on activities that should be discouraged, like pollution, not on those that should be encouraged, like employment, investment and income generation (Repetto et al, 1992). 'Tax what you burn, not what you earn', as the Alliance for Climate Protection puts it. The Federal Government requires tax revenues to pay for defense, for social security and Medicare, to pay interest due on the national debt, and for other national programs. Conservatives may grumble about government spending, but try cutting their favorite programs, like agricultural subsidies or border security or Congressional earmarks. The argument over taxes should focus on what we tax, not just on how much we tax. In order to deal with the federal budget deficit and the costs of entitlement programs, it is highly likely that at some future point tax revenues will have to be increased. Most economists agree that raising revenues through a carbon tax would be better for the economy than raising income or payroll taxes (Fullerton et al, 2008).

The real flaw in the carbon tax alternative is that nobody can tell what the tax rate should be. In the first place, it would not be a single rate but rather a series of rates rising over time. Adjusting those rates year after year in a highly contentious political process would be almost impossible. Estimates derived from leading economic models of the price needed by 2030 range from $20 per ton of carbon dioxide equivalents to more than $150 per ton and estimates for 2050 range from $60 to $250 per ton (Fawcett et al, 2009).

Why are these price estimates so divergent? First of all, uncertainty surrounds the availability of important technological options, such as new nuclear power plants, and the feasibility of sequestering large volumes of carbon dioxide from coal-fired power plants underground. A second important source of uncertainty is whether energy and power companies will be allowed to reduce their emissions indirectly by paying for lower cost reductions in other countries or through carbon-conserving changes in agricultural and forestry operations in the US. A third uncertainty concerns the future of energy prices. Should market prices be high because of rising supply costs and growing energy demand in rapidly growing developing countries, less additional economic incentive will be needed to induce higher energy efficiency and a transition to non-fossil fuels. Should future market prices for fossil energy be low, greater reliance on an explicit price for carbon emissions would be needed.

These uncertainties do not burden the upstream cap and trade approach, once it is determined what the trajectory of emission reductions over time should be. That trajectory is determined by the need to stabilize emissions at a tolerable level. Of course, the costs of emissions reductions help in determining what a tolerable level is, but – as the next chapter shows – those costs will be small relative to the damages avoided. In any case, for both bad reasons and good ones, the carbon tax proposal is not one that will go forward.

An upstream cap and trade program is simple to explain to the electorate, because it deals directly with the problem at its source: fossil fuels are the source of carbon emissions and to reduce those emissions fossil fuel use must be reduced. The policy does that. Its broader effects, however, work through a chain of cost and price adjustments by downstream industries. Unlike a carbon tax, the cap's effects on prices of various commodities and services will be brought about through market processes, not a government edict that would quickly become a political target. In that respect, an upstream cap and trade system will have political implications more like those of environmental regulations, which also generally result in higher costs and prices, but which are strongly supported politically, in part because the price effects are not readily perceptible. To put in another way, by moving the environmental restrictions far upstream, the government largely removes its fingerprints from the resulting downstream price effects. In this way, an upstream cap and trade system differs dramatically from a carbon tax in the political sphere even though its economic effects are similar.

A more plausible alternative to this ideal approach is a limited cap and trade system applied only to large power plants and perhaps some fuel-intensive heavy industries, supplemented by a package of energy efficiency standards, tax incentives for renewable energy and government support for high priority energy research. The best that can be said of this alternative is that it would be better than doing nothing. Emissions from electricity generation are only about 40 per cent of the total, so a limited cap and trade system would provide less assurance that an overall mitigation target could be achieved.

A limited cap and trade system would require substantial cuts in power plant emissions but would provide no such incentives to the majority of fuel uses and users, ensuring a pattern of abatement that is much less cost-effective. For any overall mitigation target, an upstream system that induces some reduction in carbon fuel use and greenhouse gas emissions broadly throughout all sectors of the economy will have better economic impacts than one that concentrates reductions on only a few sectors, leaving others relatively unaffected. With a broader base of coverage, the resulting energy price increases needed to achieve the overall reduction will be lower, causing less disruption and economic burden. Concentrating a cap and trade system only on electricity would have some perverse effects as well. Sharply higher electricity rates would slow down the shift in the transportation fleet toward plug-in hybrids and electric vehicles, sustaining the market for gasoline and diesel. This is undoubtedly a reason why the politically powerful petroleum industry has lobbied to be exempt from any cap and trade system.

Other policy components in the package also have limitations. Fuel efficiency standards, such as CAFE, only apply to new equipment, not to the much larger stock of existing equipment. They perversely encourage people to keep the older equipment in operation longer rather than buy new models with a higher initial cost; and, if the new models are cheaper to operate, they encourage people to use them more intensively.

Tax incentives are attractive to lawmakers, because they appear completely benign, but, of course, they have to be paid for either by higher taxes elsewhere in the code or by higher interest payments on government borrowing. They are also of little use to start-up companies with few, if any, taxable profits. Government support for energy research involves the government in 'picking winners', which in the past has often been unsuccessful – recall the high profile support for hydrogen fuel cell vehicles. It also represents a technology push approach, which does not lead to rapid penetration even of good new innovations in the absence of strong market incentives.

Finally, if Congress adopts only a limited cap and trade system, other countries will interpret the policy correctly as a limited commitment to the global goal of stabilizing the climate and will calibrate their own actions accordingly. It will be seen internationally as lack of leadership and weak participation by a US that casts itself as a world leader.

DESIGN ISSUES IN AN UPSTREAM CAP AND TRADE SYSTEM

Even if Congress enacts something like the best system just described, there will still be many important decisions to be made on issues that could potentially reduce its effectiveness and raise its costs. Many of these issues have arisen in the Waxman–Markey bill that passed the House of Representatives and in other legislative proposals.

Targets and timetables

The timing and trajectory of emissions reduction should be calibrated toward a long term stabilization goal that balances the risks of damages from climate change against the feasibility and cost of a rapid transformation of the energy system. Since there is significant uncertainty about both, the targets must have flexibility to adapt to new information about both abatement costs and climate change risks. Recent assessments have found that the risks of serious damages from climate change rise rapidly as concentrations rise from 450 parts per million of carbon dioxide equivalents toward 550. With that in mind, legislative proposals have adopted a goal of an 80 per cent reduction in emissions below 2005 levels by 2050.

Moreover, since achieving any stabilization goal requires international cooperation by major emitters in Annex I and non-Annex I countries, flexibility is also needed so that the US can participate effectively in international negotiations, making more aggressive targets in the US contingent on similar or matching actions in other key countries.

Even so, politicians are tempted to shift most of those reductions and the resulting increases in energy prices to future decades, when current office holders will probably be retired. That would be a serious mistake. As Chapter 1 explained, we may be at risk of unleashing an uncontrollable vicious cycle of global warming. Those risks would be exacerbated if emissions were to continue for another decade or two at levels nearly as high as today's.

Other consequences of delay can be seen in the European Trading System, which generously created emission permits during its first five years, thereby keeping permit prices low and discouraging electric utilities from switching from coal to gas. A better trajectory would bring about higher energy prices in the earlier decades in order to stimulate investment in new technologies. The costs of new technologies typically decline by about 20 per cent for each doubling of installed capacity, so higher initial permit prices that moderate over time will help new technologies establish themselves in the marketplace. There is little need to worry that ambitious near term targets would be unattainable. Chapter 2 showed that much could be done even with today's technologies and without heavy new investments to reduce emissions, largely through cost-effective energy efficiency improvements and behavioral changes.

Duration and banking of permits

Permits should be valid for a five-year period, which would balance energy companies' need for sufficient stability to allow them to plan ahead against the need for sufficient flexibility to allow the government to reduce permit availability as required to follow a stabilization trajectory, respond to new information and engage in international negotiations. Permits should be bankable across five-year periods, allowing unused permits to be used in subsequent periods, in order to allow energy markets to respond to economic fluctuations and to allow energy companies to plan their operations efficiently. Banking will also increase liquidity in the permit market, reduce price volatility, and attract capital from investors wishing to use the permit market for hedging or speculation.

EXTENDING THE CAP AND TRADE SYSTEM TO OTHER GREENHOUSE GASES, CARBON CAPTURE AND SEQUESTRATION

Other powerful greenhouse gases, such as methane, nitrous oxide and the halofluorocarbon gases, should be brought into the control system, because they form a significant, growing component of US emissions and some of these emissions can be controlled very cost-effectively. As mentioned earlier, emissions from large, easily monitored sources can be issued permits directly. Establishing caps for many small, dispersed sources is more difficult. Therefore, incentives to control them should be created through an offset mechanism: documented reductions in emissions of these greenhouse gases should be rewarded with tradable permits based on the equivalent carbon dioxide warming potential. Emission sources subject to the cap and trade system can then buy those permits to 'offset' their own sales of fossil fuels.

It is also important to provide similar incentives for carbon capture and storage, which is the only way for coal-fired power to remain viable, and for other forms of carbon sequestration. An offset mechanism is again the appropriate method to provide incentives. Documented sequestration of carbon, either through carbon capture and storage or through changes in land use, can be rewarded with tradable permits denominated in tons of carbon. Selling these permits would provide the economic reward for sequestration activities.

However, the same issues of permanence of sequestration and monitoring would arise as in other policy approaches. In the case of sequestration through land use changes, the danger is in rewarding activities that are taking place anyway and that contribute no additional reduction in emissions. For example, driving through the Farm Belt one sees mile after mile of fields being worked with no-till agriculture, which preserves soil carbon and reduces emissions from mechanized cultivation. The farm lobby argues that no-till agriculture should be rewarded with tradable permits for carbon capture, but since it is taking place already, such rewards would bring about no additional emission reductions. Similarly, improving manure management to control methane releases is already required by air and water quality regulations, so awarding permits for this activity would pay farmers to do what they are already required to do.

LINKING TO OTHER SYSTEMS

A national system should supersede other domestic cap and trade systems established at the State or regional level in order to avoid duplication and conflicting requirements, targets and timetables. It will not be appropriate to have a national cap and trade system operating in tandem with regional cap and trade systems operating with different rules. A national system will confront national companies with fewer compliance burdens and will obviate problems of inter-regional leakage that geographically limited systems would have to face, which is one reason why some large companies have announced their support.

A national upstream cap and trade system could link easily with cap and trade programs in other countries. Since permits will be denominated in tons of carbon equivalent, they can be fully tradable internationally. Permit holders in the US would be able to augment their domestic holdings with permits purchased abroad or sell permits into international markets. Similarly, since the US regime will include offset mechanisms, permit holders would be able to participate in international offset markets, purchasing certified emissions reductions generated through an improved Clean Development Mechanism. This will greatly increase the overall cost-effectiveness of the regime by allowing the US to stimulate and take advantage of low cost abatement opportunities in non-Annex I countries. Offsets will help reduce permit price increases and economic impacts. Chapter 6 explains how the Clean Development Mechanism can be made more reliable and larger in scale.

DOING WITHOUT A PRICE CAP

Energy industry lobbyists, supported by some economists (Philibert, 2008), have argued in favor of mechanisms to keep permit prices from rising too far or too rapidly. One such 'escape valve' mechanism would be created by government sales of additional permits into the permit market whenever permit prices reach a predetermined ceiling. It is notable that these advocates are comfortable with price volatility in underlying oil and gas markets, in which prices fluctuate by 200 to 300 per cent over business cycles. The motivation of energy interests is really to keep permit prices low, thereby allowing them to keep on with business as usual. An escape valve mechanism would weaken the effectiveness of the system. Environmental groups have successfully advocated that any permits that the government sells to stabilize prices must be borrowed from the amounts available in future years. This maintains the integrity of the overall emission reduction target but further displaces those reductions to future years, raising risks of unstoppable climate change.

Price stability can be maintained instead by linking the permit market to offset and international markets and by allowing permit banking. In addition, revising reduction targets every five years could limit undue price escalation. Energy companies will maintain reserves of permits as backing for forecasted sales. Financial institutions will also maintain reserves to support positions in the permit market. These reserves will limit price fluctuations. In addition, the cap and trade system will be linked to a much larger domestic and international market, consisting of domestic offsets generated by carbon sequestration and abatement of other greenhouse gases, certified emission reduction credits generated by the Clean Development Mechanism, and carbon permits available in the European Union Emission Trading System and in other national markets. The size and diversity of these markets will increase liquidity and limit price fluctuations.

If a price cap or escape valve were adopted, it would make linkage to international carbon markets difficult, if not impossible. If the price ceiling were set at a level above international permit prices, the ceiling would be ineffective, because US permit holders would buy permits internationally at the lower price. However, if the price ceiling were set lower than the international price, then a mechanism would be needed to prevent US traders from buying permits from the government for resale in Europe or in another foreign market. Such a trading ban would undermine the market. It is far better to rely on international markets and banking to maintain price stability.

DEALING WITH 'COMPETITIVENESS' ISSUES

Obviously, the policy approach described above would benefit some US businesses, especially those engaged in producing renewable and low carbon fuels systems, improved equipment, appliances and controls for energy efficiency, design and construction of 'green' buildings, and the like. These 'cleantech' sectors are already growing at rapid rates, attracting a flood of capital investments and creating a lot of promising, well-paying jobs.

Nonetheless, despite very little empirical support, other business interests vigorously predict the loss of industrial competitiveness if the US adopts policies to force reductions in carbon emissions. They have made similar intimidating predictions over past decades regarding enactment of all other major pieces of environmental legislation. Yet numerous studies have found negligible impacts on patterns of trade, investment or industrial location from international differences in environmental standards (Repetto, 1995). The explanation is simple: trade and investment flows are much more greatly influenced by differences in labor costs, differential access to raw materials and natural resources, and the need for proximity to growing markets than those flows are influenced by differentials in environmental compliance costs, which usually have small implications for overall production costs.

The effect of a national cap and trade system on international trade and investment would be small. The most important reason for this is that the sectors most affected by such controls do not enter significantly into international trade. Electric power is produced domestically; very little is imported or exported across US borders. Transportation services are produced domestically. Vehicles are traded internationally (although most 'foreign' automakers have domestic US production facilities), but the vehicles sold in the US are driven and emit carbon dioxide in the US. Buildings, which account for another large share of energy consumption, generally stay where they are. Organizations that establish office or commercial facilities in other countries generally do so for other reasons than to save on energy costs. Government services, which represent another 20 per cent of the economy, generally do not lose their share of the US market to other governments with lower energy costs. Wholesale and retail trade and a host of other services are anchored to the consumer and the nearest shopping malls.

Even within the industrial sector, energy costs represent a small percentage of total production costs, well below five per cent, in most industries. In most of the dynamic, technologically advanced manufacturing and service industries in which the US has a comparative advantage, energy costs represent an even lower share in production costs. In those sectors, international differences in energy costs make up only a fraction of that small percentage, not enough to affect trade or investment decisions significantly. Only for a subset of heavy industries, such as chemicals, metals, cement and other nonmetallic minerals, are energy costs really significant. And for some of these, such as cement, high transportation costs relative to value make production largely a national affair.

Even in these sectors, another important reason to discount the threat of competitive impacts is that other countries, including our most important trading and investment partners in Europe, Canada and Japan, have already adopted or agreed to adopt their own mandatory limits on carbon emissions. Among the Annex I countries, which account for the large majority of international trade and investment flows, it is only the US that has not adopted mandatory limitations. Fuel prices and electricity rates in countries that have already adopted mandatory carbon controls tend to be higher, and sometimes considerably higher, than they are in the US. That being so, if competitiveness impacts were indeed important, one would have expected to see a flight of industry to this country. Moreover, the European Union has already made clear that it is prepared to order steeper cuts in the next phase following 2012 if the US adopts mandatory limits. China and India have also taken actions to reduce emissions and have signaled that their future actions are conditional on actions taken in the US and other rich countries.

Competitiveness fears are misplaced. Some legislative proposals contain provisions requiring protectionist tariffs against imports from countries that US agencies deem not to have policies equivalent to ours. Not only would that be contrary to international trade agreements administered through the WTO, it is also a two-edged sword that could be turned against the US by countries that have adopted stricter emission controls. It is a policy approach easily subject to abuse, as the 'dumping' provisions in trade law have been abused for protectionist purposes. The victims are not only foreign producers, but also US companies that would have to pay higher prices for steel, aluminum, chemicals and other industrial inputs that are protected in this way.

Nonetheless, it is important for the US, along with other Annex I countries, to negotiate with large non-Annex I countries such as China, India, Mexico and Brazil to ensure that they participate and cooperate significantly in the next phase of greenhouse gas emission reduction. Emissions from those countries are large and growing. Over time, unless they are reduced, it will be impossible to achieve global climate stabilization at any relatively safe level. That fact, rather than fears of competitive impacts, should underlie negotiations.

DEALING WITH 'FAIRNESS' AND EQUITY ISSUES

National climate policies will have significant and long-lasting economic implications, so concerns about fairness are understandable. An upstream cap and trade system that keeps economic impacts low and spreads them broadly throughout the economy by price increases in all fossil fuels is fairer than one that concentrates reductions and impacts on a smaller segment of the economy. In an upstream system, every household will bear part of the burden in proportion to its direct and indirect use of fossil fuels. Those more affluent households that use a lot of energy, one way or another, will bear a larger share of the burden. That's only fair.

Still, energy bills represent a slightly higher percentage of monthly expenditures for lower income households than for upper income households. The potential impacts of higher energy costs on low income households are therefore a legitimate concern (Rausch et al, 2010). These poverty-related effects can easily be dealt with, but there are better ways and worse ways to do so. The worst way would be to prevent retail energy prices in the residential sector from rising. Doing so would blunt people's incentive to increase energy efficiency in their homes, even by such simple steps as turning off lights and appliances when they're not in use, applying insulation around windows and doors, or adjusting thermostats. A better way would be to compensate low income households with additional income payments not tied to their energy usage. This can be done in several ways: by cost of living adjustments to social security, earned income tax credit and other social welfare programs; by lump sum rebates delivered through electricity distribution companies; and by directly targeted government lump sum payments to low income households.

Other appeals to 'fairness' are much less compelling. It is ludicrous to behold the CEO of a major electric utility that regularly appears before State public utility commissions arguing for the biggest rate increases it can get pleading for billions of dollars worth of free permit allowances so that he will not have to raise electricity rates on poor widowed grandmothers with fixed incomes. Retail sales are typically only a third to a half of utility sales and only a tiny fraction of retail sales are to poor widowed grandmothers. Many States already have programs to help low income consumers with energy bills. It is better to compensate Granny directly than to give away billions of dollars in free permits to large energy companies.

Fairness can be improved by auctioning off most of the permits. If permits were all distributed free to sellers of fossil fuels, a process known as 'grandfathering', they would receive from government a very valuable, salable asset. Annual carbon emissions in the US are more than 5 billion tons. If the initial carbon permit price established through trading were $30 per ton, then oil, gas and coal companies would receive a windfall on their balance sheets of $150 billion. This asset gain would be reflected in their stock prices, as experience in the European Emission Trading System has shown (Point Carbon, 2008). The beneficiaries would ultimately be their shareholders, a relatively wealthy group, while the costs would be borne by energy users (Parry, 2004). If a substantial fraction of the permits are auctioned off, some of the resulting revenues can be used to cushion impacts on relatively vulnerable households, through any one of the several fiscal measures mentioned above, and to compensate them for higher energy costs.

Another significant portion of revenues derived from a permit auction can be used to reduce other tax rates or to reduce the federal deficit and forestall future tax increases. Recycling auction revenues in this way will also reduce the burden on households or businesses and will significantly limit the overall economic impact of the regime by stimulating household consumption and labor supply and business investment, depending on the specific tax offsets chosen. Many economic studies have concluded that recycling auction revenues through offsetting tax cuts markedly increases the cost-effectiveness of a cap and trade program (Cramton and Kerr, 1998).

Though permits to emit sulfur and nitrogen oxides were grandfathered to electricity generators in cap and trade programs under the Clean Air Act, a cap and trade program for carbon dioxide would have substantially different implications. First of all, carbon dioxide emissions are far greater and the total value of carbon permits would also be much greater, so grandfathering would represent a larger windfall for carbon permit holders. Also, since the sulfur and nitrogen cap and trade programs required much larger percentage cuts in emissions than the carbon program would require in its early decades, there were fewer sulfur and nitrogen permits left on the companies' books than carbon permit holders would have.

It is worth remembering that the atmosphere is a vital public resource. Companies that use other public resources pay for the privilege: oil leases on public lands and in federal waters are auctioned off and oil companies pay royalties to the Federal Government on oil and gas they extract; companies that cut timber or graze cattle in national forests pay fees to the Federal Government; companies that use the electromagnetic spectrum to broadcast their messages must purchase rights for the frequencies they use at auction. Why should energy companies freely use the atmosphere as a dump for their exhausts?

It would only be necessary to grandfather a small fraction of permits to oil, gas and coal companies to compensate them fully for their losses in sales (Goulder and Bovenberg, 2005). The reason for this is that an upstream cap and trade company would be equivalent to a government-enforced cartel for those industries. OPEC is a cartel. Its members agree to limit production and sales of petroleum in order to maintain higher prices. Its weakness is that members cheat by surreptitiously selling more petroleum than their agreed quotas, and there is nothing that other members can do about it except sell even more oil themselves to drive down the price. Like a cartel, an upstream cap and trade regime would also limit sales and drive up prices in the US market. The difference is that the government would enforce the quotas and prevent cheating. The cartel would be more effective. The higher prices would largely or entirely compensate the sellers for the loss in sales, so little further compensation should be required in the form of free permits.

Another unpersuasive claim to 'fairness' arises from States that derive most of their electricity from coal-fired power plants. They claim that they would be disproportionately burdened, because their electricity prices would rise most, since coal is the most carbon-intensive fuel. Those States have and have had some of the lowest electricity prices in the country, not only because coal has been inexpensive, but also because they have hosted some of the country's most polluting power plants, producing emissions that drift across State borders.[2] Table 3.1 contrasts the States with some of the largest and smallest percentages of electricity generated from coal. On both the East and West Coasts, where those percentages are small, retail power prices are twice as high as in the interior States highly dependent on coal.[3] The exception is Washington State, which is blessed with abundant hydroelectric power.

Moreover, those coal-dependent States have long been the beneficiaries of federal largesse. The Tax Foundation computes the ratio

Table 3.1 Most and least coal-dependent States

State	Per cent electricity from coal	Average retail electricity price (cents/kWh)	Federal spending/tax ratio	Per capital sulfur dioxide emission (tons)
Washington	6	6.55	0.88	3.4
California	1	12.48	0.78	0.03
New York	15.1	16.57	0.79	9.8
Connecticut	12.4	17.79	0.69	2.9
Massachusetts	2.5	16.27	0.82	13.1
Utah	89.7	6.49	1.07	12.5
North Dakota	93.5	6.69	1.68	212.4
Kentucky	92.4	6.26	1.52	116.5
West Virginia	97.7	5.61	1.76	257.9
Indiana	94.9	7.09	1.05	136.4

Source: Energy Information Agency, US Department of Energy

between federal spending in a State and federal taxes collected from individuals and businesses in it (Tax Foundation, 2005). Those States most dependent on coal typically receive much more in federal spending than they pay in federal taxes.[4] The coastal States least dependent on coal typically receive much less in federal spending than they pay in taxes. It is unclear why coal-dependent States should receive even more consideration in national policy.

THE COSTS OF CHOOSING AN INFERIOR POLICY APPROACH

The advantages of an upstream cap and trade system like the one this chapter describes have been recognized by public and private policy research groups, including the Congressional Budget Office, Resources for the Future and the Climate Policy Center. The National Commission on Energy Policy has also endorsed a comprehensive nationwide cap and trade program, implying an upstream approach. Unfortunately, most of the recently proposed legislation and the Waxman–Markey bill that passed the House of Representatives depart significantly from this approach.

It is extremely important that federal policies adopted in the US be both effective in reducing emissions and cost-effective in keeping the resulting economic impacts to a minimum. There is a significant danger that policies adopted through political negotiations will prove to be inferior. They may lead to unintended consequences, undermining the policy objectives. They may give rise to substantial costs in excess of those required by a more cost-effective approach. They may do both.

The policy regime previously adopted by the European Union illustrates the point. The limited coverage of the European Emission Trading System left out many emissions sources that might have been controlled at relatively low cost. The absence of provisions to allow banking of permits created price volatility in the carbon market. The over-allocation of permits when the program began undermined incentives for firms to reduce emissions. The grandfathering of permits to such firms provided a windfall on their balance sheets while doing nothing to offset or cushion price increases to consumers.

Policies to control emissions will require an energy transition with economic implications for all households, industries and regions to a greater or lesser degree. As explained in the next chapter, almost every macroeconomic model that has been used to study the issue finds that such impacts need not be large if a cost-effective policy approach is adopted. These economic analyses find that the impact, at worst, would represent a slight reduction in the rate of economic growth, which would be more than offset by the reduction in environmental damages. However, these models also find that the difference in economic costs between a cost-effective policy and an inferior approach could grow to as much as one per cent of gross domestic product per year. One per cent of GDP is about $150 billion, a high price to pay for misguided policy choices.

What is more worrying is that policies to control greenhouse gas emissions must be kept in place at least for decades and perhaps for a century. Over a decade, the excess cost of an inferior policy choice could be $1.75 trillion or more, given the annual growth in the economy. This would be enough to resolve the social security shortfall, fund expanded healthcare coverage, eliminate the budget deficit or fulfill many other worthy public goals.

It is crucial that such mistakes be avoided. If mistakes are made, it is highly unlikely that they will be corrected by switching later on to a better policy architecture. The excess costs, then, will go on indefinitely, piling up huge economic bills decade after decade. Policies tended to get 'locked-in', as leading political scientists have noted. Policy choices are path-dependent: choices made at the outset constrain the options available later on and raise the costs of switching to a different policy regime (Coglianese and D'Ambrosio, 2008). Policy lock-in occurs for several reasons:

- When a policy is adopted, specific institutional and administrative investments are made to support it, both in government and in the private sector. Such investments and start-up costs are written off very reluctantly.

- When a new policy is adopted, considerable policy learning takes place, as those involved learn to operate within that framework. That learning gives a familiar regime both cost and psychological advantages over a new and unfamiliar approach.

- Most importantly, policies engender interests that benefit from the specific rules adopted. These may include government bureaucracies, politicians able to allocate benefits and costs, firms that have made investment commitments with profits contingent on the policy continuing, as well as community and other interest advocacy groups that benefit from the policy. All these interests can be expected to lobby against policy change.

Because the excess costs of inferior policy choices are so high, because those costs will continue and grow over such a long period of time, and because it will be so difficult to change to a superior approach if an inferior policy is first adopted, it is of critical importance to get it right the first time – to adopt a policy architecture that is both effective and cost-effective, even if political expediency leads in another direction.

NOTES

1   In the US, a small fraction of energy products is exported or used as feedstocks in the chemical industry. These sales will be tracked and credited against the sellers' permit accounts. A mechanism will be needed to ensure that sales to industrial companies outside the permit systems are not used as a conduit to resell products for fuels, circumventing the permit system.

2   For the lower 48 States, the correlation coefficient across States between the percentage of electricity derived from coal and per capital sulfur dioxide emissions is 0.66.

3   Using data collected from the Energy Information Agency of the Department of Energy, the correlation coefficient across States between the percentage of electricity derived from coal and average retail electricity prices in the lower 48 States is −0.60.

4   Across the lower 48 States, the correlation coefficient between the ratio of federal spending to federal taxes and the percentage of electricity derived from coal is 0.32.

REFERENCES

Broder, J. (2009) 'House bill for a carbon tax to cut emissions faces a steep climb', *New York Times*, 6 March, pA13

Coglianese, C. and D'Ambrosio, J. (2008) 'Policymaking under pressure: The perils of an incremental response to climate change', *Connecticut Law Review*, vol 40, p41

Congressional Budget Office (2008) *Policy Options for Reducing CO$_2$ Emissions*, Washington, DC

Cramton, P. and Kerr, S. (1998) *Tradable Carbon Permit Auctions: How and Why to Auction not Grandfather*, Resources for the Future, Washington, DC

Fawcett, A., Calvin, K., de la Chesnaye, F., Reilly, J. and Weyant, J. (2009) 'Overview of energy modeling, Forum 22 US Transition Scenarios', *Energy Economics*, vol 31, ppS198–S211, http://emf.stanford.edu/files/res/2369/fawcettOverview22.pdf

Fullerton, D., Leicester, A. and Smith, S. (2008) *Environmental Taxes*, National Bureau of Economic Research, Cambridge, MA

Goulder, L. and Bovenberg, A. (2005) 'Efficiency costs of meeting industry distributional constraints under environmental permits and taxes', *Rand Journal of Economics*, vol 36, pp151–164

Hahn, R. and Hester, G. (1989) 'Marketable permits: Lessons for theory and practice', *Ecology Law Quarterly*, vol 16, pp361–406

Inter-agency Task Force on Carbon Capture and Storage (2010) *Executive Summary: Report of the Inter-agency Task Force on Carbon Capture and Storage*, Washington DC, August

Parry, I. (2004) 'Are emission permits regressive?', *Journal of Environmental Economics and Management*, vol 47, pp364–387

Philibert, C. (2008) *Price Caps and Price Floors in Climate Policy: A Quantitative Assessment*, International Energy Agency, Paris, accessed August 2010 at www.iea.org/papers/2008/price_caps_floors_web.pdf

Point Carbon Advisory Services (2008) *EU ETS Phase II: The Potential and Scale of Windfall Profits in the Power Sector*, Report for WWF, London, accessed August 2010 at www.wwf.org.uk/filelibrary/pdf/ets_windfall_report_0408.pdf

Rausch, S., Metcalf, G., Reilly, J. and Paltsev, S. (2010) 'Distributional implications of alternative US greenhouse gas control measures', *The B.E. Journal of Economic Analysis & Policy*, vol 10, Issue 2 (Symposium), Article 1, Berkeley, CA, accessed August 2010 at www.bepress.com/bejeap/vol10/iss2/art1

Regional Greenhouse Gas Initiative (2007) 'Overview of RGGI CO_2 Budget Trading Program', http://rggi.org/docs/program_summary_10_07.pdf

Repetto, R. (1995) *Jobs, Competitiveness and Environmental Regulation: What are the Real Issues?*, World Resources Institute, Washington, DC

Repetto, R., Dower, R., Jenkins, R. and Geoghegan, J. (1992) *Green Fees: How a Tax Shift Can Work for the Environment and the Economy*, World Resources Institute, Washington, DC

Tax Foundation (2005) 'Federal spending received per dollar of taxes paid, by State', accessed August 2010 at www.taxfoundation.org/taxdata/show/266.html

Tietenberg, T. (2006) *Environmental and Natural Resource Economics* (7th edition), Addison-Wesley, New York

US Department of Energy (2009) *Emissions of Greenhouse Gases in the US 2008*, Energy Information Agency, Washington, DC

US Environmental Protection Agency (2010) http://epa.gov/methane/scientific.html

Wang, M. (2010) 'Government subsidizes deep-water drilling', *ProPublica*, accessed 25 May 2010 at http://propublica.org/blog/item/oil-companies-still-get-billions-in-incentives-to-drill-in-deep-water

Western Climate Initiative (2009) 'Design recommendations for the Western Climate Initiative Regional Cap and Trade Program', accessed August 2010 at www.westernclimateinitiative.org/component/remository/general/design-recommendations/Design-Recommendations-for-the-WCI-Regional-Cap-and-Trade-Program

Woody, T. (2010) 'Clean technology investing slips but could be worse, report finds', *New York Times* Green Blog, 6 January

Chapter 4

The New Economics and Climate Policy

The preceding chapters have described the energy transition needed to prevent catastrophic climate change and the most cost-effective policies needed to bring this transition about. Legitimate concerns have been raised about the economic consequences of these policies: their impacts on energy prices, on employment, on consumers, on industries and on prospects for future gains in living standards. What is known about these issues?

TWO COMPETING REPRESENTATIONS OF THE ECONOMY

Predicting the economic consequences and impacts of this energy transition juxtaposes two very different representations of the economy. In one view, the economy evolves through technological innovation, the creation of new products, and behavioral responses to new technological opportunities and resource constraints. Economic growth and improvements in living standards emerge not primarily from the replication of existing production and consumption patterns but from dynamic change. In search of the profits potentially available by exploiting new technologies and markets, entrepreneurs and investors set off cascades of innovation and growth. New firms and industries emerge and grow; others decline. At any moment, some firms and households are on the leading edge of technological change; others are lagging behind. Innovation, entrepreneurship and investment respond to the opportunities presented in ways that cannot now be fully foreseen.

Providing empirical confirmation for this representation of the economy, some of America's best minds are already responding to the challenge of climate change and are heavily engaged in efforts to develop better batteries, fuel cells, solar photovoltaics, algal-based biofuels and many other energy innovations, with uncertain results. Investments have been streaming into venture capital funds that hope to profit from these emerging market opportunities. With a supportive policy framework in place, some of these efforts may give rise to large, rapidly growing, profitable new industries that become the drivers of economic growth. Other efforts may fail for technological or commercial reasons. Picking the winners in advance is difficult.

This evolutionary view of the economy (Arthur, 2009) suggests that the transition to low carbon fuels could set off a surge of economic innovation and growth comparable to previous energy transitions: the transition from rail and animal transport to petroleum-based motor transport, for example, or the transition from steam power to electricity. This evolutionary view of economic history draws on Josef Schumpeter's famous description of the process of 'creative destruction' in capitalist economies (Schumpeter, 1945). New industries grow; old industries adapt or perish. This process can be seen in the pervasive effects of the semiconductor and the computer in the American economy, enormously stimulating productivity and innovation, but relegating many earlier products and industries to extinction.

Although this representation better reflects the dynamism of the American economy, in which technological change is rapid and possibly accelerating, it has not been used to analyze the economic implications of mitigating greenhouse gas emissions. Evolutionary economics does not lend itself to analyses that yield definite forecasts or predictions. The 'open' characterization of the economy, recognizing that new conditions and capabilities emerge over time, inevitably makes predictions uncertain and indeterminate. Will genetic engineering enable massive and cost-effective production of biofuels? One cannot foresee with any confidence, any more than Thomas Watson, the founder of IBM, could foresee all the applications of digital computing. Will better means of energy storage be developed that will permit much greater deployment of intermittent wind and solar energy resources? One cannot foresee any more clearly than Thomas Edison could foresee the development of the immense variety of electric appliances.

In the other representation, which characterizes all the economic models used to analyze the consequences of greenhouse mitigation, all future technologies, capabilities, resource availabilities and prices are assumed to be known in advance. Firms and households respond to these foreseen conditions by developing optimal strategies that maximize profits or welfare. The set of possibilities included in the analysis must be 'closed' and limited in order that a maximizing strategy can be found. The set of technological and other possibilities available not only today but in future decades must be assumed known, both to the analyst creating the model and to the economic agents – firms and households – represented in it. In some models, the set of technologies represented in the model is extensive, though still closed. In some models, predetermined improvements in costs or productivity over time are programmed into the model, but are still assumed from the outset. In these models, the economy quickly achieves equilibrium based on efficient deployment of resources under assumed constraints and proceeds along a trajectory that allows all firms and households the best outcomes, given the technological possibilities that have been programmed into the analysis.

The advantage of this representation is that with sufficient further simplifications it can generate definite results representing the best outcomes.[1] The implications of various policy choices can then be compared to each other and to a 'business as usual' scenario that assumes no deliberate policy effort to mitigate greenhouse gases. Models can be run repeatedly under different assumptions regarding policy design, technological and resource availabilities, and costs, in order to explore the impact of variations in these important conditions. Nonetheless, all such analyses harbor the key assumption that everything is known today about the technological opportunities and conditions that will emerge over the coming three or four decades – a heroic assumption.

In assuming that all households and companies are in a welfare-maximizing, profit-maximizing equilibrium position from the outset, this representation also implies, contrary to widespread empirical evidence, that there are no opportunities to reduce greenhouse gas emissions and save money while doing so. Armed with perfect information and foresight, firms and households represented in the model are assumed to exploit all available cost-saving investments and purchases within the set of available technologies. This is assumed to be true both in the policy simulations and in the baseline, 'business as usual' simulation. If, in these analyses, the damages from climate change are set aside and the focus is only on mitigation costs, the implication must be that *any* deviation from the business as usual scenario must raise costs and reduce welfare, since in the business as usual case the economy was assumed to be already operating as efficiently as possible.

In stark contrast, many empirical studies of mitigation possibilities available to firms, households and public sector institutions have found that a substantial percentage of total US emissions, from 20 to 30 per cent, can be eliminated with savings in costs or with a super-normal return on the required investment (McKinsey & Co, 2007). These potential savings can be found in buildings, appliances, industrial processes and equipment, energy conversion, and transportation (von Weizsäcker et al, 2009). Since such opportunities contradict the assumption of efficient decision-making, many economists are skeptical that they can actually exist, but attempts to understand their persistence have uncovered a variety of explanations: lack of information, organizational and personal inertia, misaligned incentives, and institutional obstacles, among others.

Overcoming these obstacles presents a broad agenda of potential policy interventions that could reduce emissions either at little or no cost or with actual savings. Furthermore, the potential savings are almost certainly understated, because studies of so-called 'win–win' opportunities have examined each one in isolation, not taking into account their cumulative effects. For example, improved insulation, better use of passive solar design and natural lighting and ventilation, and water-saving equipment might all be cost-effective in isolation, but when deployed together they might also allow down-sizing of the building's HVAC system, with an additional saving in capital investment. Should such efficiency gains be captured in many buildings on a large scale, further savings might be possible by reducing investments in the upstream supply and delivery of energy. None of these possibilities are reflected in models that assume complete energy efficiency as a baseline condition.

The assumptions of predetermined technologies and perfect efficiency in the analytical models used to assess the costs of greenhouse gas mitigation are reason enough to view their findings with caution, especially with regard to outcomes over the long term. The analyses are best suited to explore the implications of alternative assumptions regarding technological availabilities, economic reactions and policy alternatives. Nonetheless, since economy-wide policies to reduce greenhouse gas emissions have not yet been enacted, little other than these models are available on which to base predictions regarding costs and impacts.

Many such models have been constructed by university economists, public research institutes and private consulting firms to analyze the economic costs and impacts of greenhouse gas mitigation. Though they share the same basic representation of the economy, they differ considerably in structure and assumptions, and project quite different costs for the same policies and mitigation trajectories. Efforts have been made to understand how such differences arise, both by examining the models in detail and by carrying out formal 'meta-analyses' that associate differences in predictions with the differences in model assumptions

(Repetto and Austin, 1997). These efforts have identified both the most crucial assumptions that lead to different results and the commonalities in the findings of various economic model analyses.

THE BOTTOM LINE

The most significant findings from these economic investigations include the following:

+ Even under worst case assumptions built into models, as greenhouse gas emissions fell by 80 per cent by 2050 in response to an economy-wide cap and trade program, economic impacts would be mild and economic growth would continue robustly despite higher delivered energy prices.

+ Under worst case assumptions, gross domestic product and household consumption might be 1–3 per cent lower by 2030 than in the baseline scenario because of higher energy prices. This implies a rate of economic growth over two decades that is only marginally slower, from about 2.71 per cent per year to 2.68 per cent per year. This predicted difference in growth rates is much smaller than the error such models make in forecasting economic growth over these lengths of time. Under less pessimistic assumptions, the predicted impacts on economic growth are even milder (Paltsev et al, 2009).

+ The predicted impacts on household welfare are smaller still, in part because households will adjust to the higher prices of purchased goods and services by producing more of them themselves (Fawcett et al, 2009). For example, there would be fewer trips out for entertainment and more evenings watching videos and making dinner at home.

+ Even aside from the averted damages from climate change,[2] other benefits would offset a significant fraction of these economic costs.

These benefits, which are considered in only a few of the economic models, would include reduced mortality, morbidity and healthcare costs from improved air quality, and reduced dependence on imported oil (Bollen et al, 2009).

The bottom line is that greenhouse gas emissions can be drastically reduced through a transition to clean and efficient energy without significantly lowering living standards or sacrificing future economic gains. All reputable economic analyses agree on this basic finding.

That said, the availability of technological options significantly affects costs. If the expansion of nuclear power is limited, if carbon capture and storage from coal- and gas-fired power plants proves infeasible or prohibitively expensive, or if the expansion of wind, solar and geothermal power is restricted by a lack of transmission facilities, then the costs of achieving the reduction in emissions will be substantially higher.

The analyses also find that policy choices are important: good policy choices can substantially reduce costs. As the preceding chapter explained, the most cost-effective way of implementing a large, long term reduction in greenhouse gas emissions is to create an economy-wide price on carbon through a comprehensive cap and trade system. 'What, where and when' flexibility provided by market-based policies lowers costs compared to command and control approaches. A comprehensive approach lowers costs by including all sources of emissions. If some sources are left uncontrolled, some low cost mitigation options may be sacrificed and tighter controls required on the remaining ones, leading to inefficiencies, higher abatement costs and higher energy prices.

Taking advantage of all potential mitigation and sequestration options, including those in forestry and agriculture, can significantly lower costs. Similarly, allowing US firms to take advantage of low cost sequestration and mitigation options in other countries would also lower costs substantially. These basic findings are shared among all serious economic analyses.

THE RANGE OF PREDICTIONS FROM LEADING MODELS AND ANALYSES

The Energy Modeling Forum at Stanford University[3] has for many years performed a useful service by bringing together leading analysts investigating these issues and enabling them to understand why their models produce different results. Analysts run their models using the same assumed scenarios regarding the future mitigation trajectories and then investigate why the models obtain different results.

Almost all of the models included in these comparisons have been developed either by macro-economists at leading universities or by economists at the national research institutes. Their projections have figured prominently in policy discussions and continue to do so. The following figures illustrate the range of predictions these models generate. That range excludes the possibility of disastrous economic impacts predicted by some industry propagandists and politicians.

Figure 4.1 illustrates (the gray lines at the top) the projected business as usual emissions from 2000 to 2050 in six of these leading models (Fawcett et al, 2009). All projections are referenced to the latest available Annual Energy Outlook produced by the Department of Energy's Energy Information Agency and take into account the most recent available energy legislation and economic forecasts (US Department of Energy, 2010). Nonetheless, it is obvious that different models predict widely different growth rates of emissions in future years under business as usual assumptions. In the most optimistic, emissions grow from about 7 to about 8 billion tons over 50 years; in the least optimistic, they increase to about 11 billion tons, a percentage increase more than three times as rapid.

The dashed lines represent the mitigation trajectories that all analysts used as a basis for comparing model results. The steepest decline represents approximately an 80 per cent reduction from emissions in 2005, the trajectory proposed in recent Congressional legislation.

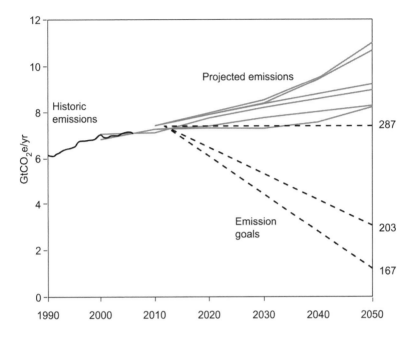

Source: US Department of Energy

Figure 4.1 *Projected baseline emissions and mitigation trajectories*

Naturally, the higher the emissions are projected to be in the business as usual scenario, the more tons must be eliminated to achieve any mitigation target, and the higher the costs are likely to be. The differences in projected baseline emissions are attributable to differing assumptions about the pace of future labor force and economic growth, the pace of ongoing improvements in energy efficiency, and the shift away from carbon fuels in the absence of further policy stimuli. These assumptions play an important role in generating differences in predicted costs and economic impacts.

The lines in Figure 4.2 are generated by six of the leading models investigated by the Energy Modeling Forum (US Environmental Protection Agency, 2009). The lower panel corresponds to the 80 per

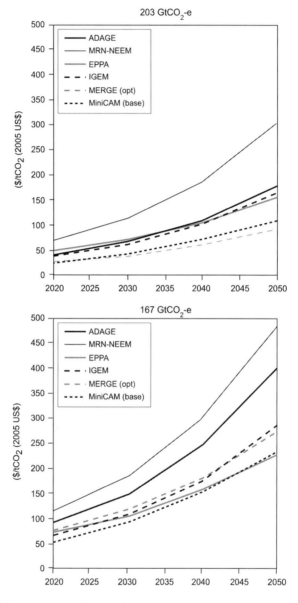

Source: US Environmental Protection Agency

Figure 4.2 *Projected carbon emission permit prices in leading models*

cent emissions reduction by 2050. All models assumed that the mitigation trajectories would be achieved by an economy-wide cap and trade regime that required all covered emission sources to hold a permit for any and all emissions during the year. The number of permits available each year would decline in step with the mitigation target. Permits would be tradable freely among sources, leading to one economy-wide permit price, which represents the minimal cost to fulfill the final ton of each year's mitigation obligation.

Although some economists have argued that the most effective policy response to the climate problem would be to put a price on carbon emissions by levying a carbon tax, Figure 4.2 illustrates why this approach would be unworkable in practice, as the previous chapter indicated. First, what would be needed would not be just one carbon tax at a fixed rate, but rather a sequence of carbon tax rates that increases year by year. It is unrealistic to think that a carbon tax, which would generate $50 to $100 billion in tax revenue each year and generate intense political controversy, could be subject to such steady rate changes. Second, there is considerable uncertainty about the tax rate that would bring about any given mitigation trajectory. In the projections produced by various analyses, there is a three- or four-fold difference in the carbon price consistent with a given mitigation target: in 2030 the required tax rate might be as low as $30 per ton or as high as $120 for the easier mitigation target. Even those favoring a carbon tax would be hard pressed to choose the appropriate rate.[4]

REASONS WHY PREDICTIONS DIFFER SO WIDELY

Future technology options

Why is the range of price uncertainty so large, even in models in which everything is assumed known at the outset? In addition to the differences illustrated in Figure 4.1 in the number of tons that would have to be abated, the assumptions about technology availability play an important role. If availability is unrestricted and the best current estimates of future costs are assumed for various energy technologies, most models would forecast that the role of nuclear power would expand substantially as a source of low cost baseload electricity. Future coal plants coming on line after 2025 would be equipped to capture carbon dioxide and sequester it underground. If assumptions are built into models that limit the availability of these options or significantly raise their estimated future costs, then the predicted permit prices and overall costs of meeting mitigation targets would be higher. More reliance would have to shift to reductions in energy demand and to the remaining low carbon energy sources.

Other assumptions about technology availability are buried deeply within model structures and require a close examination of models' documentation to be understood. For example, some models assume very limited substitutability between fossil fuel power plants and wind or solar generation, such that the share of the last two cannot rise beyond 20 per cent of power generation, no matter how high the price on carbon emissions might go. Understandably, this restriction implies much higher carbon prices, especially if combined with limitations on nuclear power and carbon sequestration. The justification for the assumption that intermittent wind and solar power can play only a minor role in the generation portfolio is that there must always be enough power instantaneously available to ensure reliable fulfillment of demands at all times.

However, if future solar and wind power installations are linked in an extensive geographic transmission grid, some fluctuations will average out. In addition, new large natural gas deposits recoverable from shale formations may encourage solar thermal and gas turbines to be paired in order to ensure economic and reliable low carbon electricity generation. Such plants are already under construction. Moreover, there may be advances in energy storage possibilities, including the use of vehicle batteries in electric vehicles connected through a smart grid.

In some models, assumptions are built in that imply rising cost curves as the installed capacity of wind and solar power expands. These assumptions are not based on historical data, which show declining costs for these renewable power technologies as installed capacity has increased. Models that project declining costs into the future understandably imply lower overall mitigation costs and permit prices.

Important technological assumptions are also built into descriptions of energy-using sectors. The degree of substitutability between energy inputs to production and the use of capital and labor is one such significant assumption. In some models, production decisions are modeled as involving first a decision of the optimal mix of capital and labor, and after that a decision about how much energy to use with this mix. That's not generally the way it works: most tradeoffs regarding energy use come at the stage of plant design, when decisions are made whether or not to invest a bit more capital to install more energy-efficient equipment, thereby lowering future operating costs. After that, managers decide how much labor to use to staff this plant, for example whether it should be operated for one daily shift or two or three. Depending on the substitution elasticities chosen, such assumptions can significantly affect energy-saving responses to price signals. Importantly, these conditions are simply assumed in most models, using parameters not estimated from historical data but chosen by the analyst.

Since the energy transition will take place over several decades, what the models assume about the drivers and pace of technological change is important. Typically, they assume a gradual improving trend in the energy efficiency of the economy, reflected in a trend of falling energy use per dollar of gross domestic product. Some models go further by extrapolating trends in the cost improvements of some energy conversion technologies. The more rapid the assumed rates of technological improvement, the more favorable the economic impacts. Very few models go even further and embody assumptions that relate the rate of improvement in these technologies to the cumulative investment made in them. The history of many industries reveals that as experience with a technology and investment in it increase, a series of incremental improvements leads to lower costs, a form of 'learning by doing'. The few models that encompass these learning curves of induced technological improvements find that economic impacts are more favorable, because the economic incentives that promote a low carbon technology lead to cost improvements over time in those same technologies.

Future energy prices

Understandably, assumptions with regard to future energy prices have important implications for the predicted costs of the energy transition. If petroleum prices are assumed to increase in future decades to historically high levels, reflecting the increasing marginal costs of petroleum extraction from difficult resources, then models predict that only a smaller carbon price would be needed to produce the transition to an electrified and fuel-efficient vehicle fleet.

Similarly, if models reflect recent findings from the US Geological Survey studies reducing previous estimates of economically recoverable coal resources (Pierce and Dennen, 2009) and therefore predict higher future coal prices, then they will conclude that the transition away from coal-fired power plants will take place at a lower carbon price. Understandably, the more accessible, lower cost coal reserves tend to be exploited before the less accessible ones, implying a tendency toward rising costs over time unless offset by sufficient technological improvements in mining techniques. Moreover, a substantial fraction of potential resources are already inaccessible due to environmental or other constraints.

Finally, to the extent that models incorporate the higher estimates of natural gas availability in future decades, reflecting potential exploitation of large volumes of gas trapped in shale deposits, and therefore predict a lower trajectory of future gas prices, then models will predict that gas-fired power plants will substitute for coal-fired power at a lower price on carbon emissions. The lower the carbon price needed to bring about the energy transition, the smaller the economic impacts. These uncertain forecasts of future fossil fuel prices have important implications for model predictions.

Policy choices

Despite their limitations and uncertainties, economic models can be useful in assessing policy choices. Such analyses and actual past experience in environmental regulation have shown that market-based policies, such as cap and trade regimes, can reduce costs. The broader the coverage of such regimes, the greater will be the number and diversity of emission sources that will have incentives to reduce their release of greenhouse gases. That's why an upstream cap and trade system is the best policy option. One study found that exempting just three sectors responsible for 17 per cent of emissions would raise permit prices by 30 per cent and welfare costs by 30–50 per cent (Paltsev et al, 2009). Limiting a cap and trade system to the electric utility industry, which contributes just 40 per cent of US carbon emissions, would substantially raise the costs of achieving any national emission reduction target.

Partial and piecemeal policies are less inefficient and raise costs. For example, vehicle fuel efficiency standards reduce emissions per mile but fail to encourage owners to drive fewer miles. Perversely, because operating costs are reduced, drivers tend to drive more miles, not fewer. Similarly, policies that shield electricity utility customers from rate increases that reflect the price of carbon blunt users' incentives to conserve electricity, forcing more emission reductions to be found elsewhere.

Nonetheless, 'market fundamentalism' is unwarranted. Just pricing carbon is not the most effective policy approach. Complementary institutional changes, information programs, well-designed regulations and enabling infrastructure investment like those described in Chapter 2 will reinforce market incentives and smooth the way for an energy transition. Using such measures, California managed to keep electricity use level for decades despite population and economic growth.

An upstream cap and trade system will cover almost all carbon dioxide emissions. There are also cost-effective ways to reduce levels of other powerful greenhouse gases, such as methane, nitrous oxides and certain industrial gases, and there are carbon sequestration opportunities in soils and forests. Allowing users of fossil fuels to reduce their net emissions by contracting with sources of other non-carbon greenhouse gases to carry out abatement measures, or with those able to sequester carbon out of the atmosphere, or with emitting sources outside the US is one of the principal means to reduce costs. Analyses agree that if such offset programs are restricted, as some legislative proposals have done, significant low-cost options to reduce emissions or to enhance carbon sequestration will be passed over. Consequently, the costs of achieving mitigation targets will be significantly higher.

Questions have been raised, based on past experience, whether such offset programs have really contributed to a global reduction in emissions or whether they have simply rewarded measures that would have been taken even without any further incentive, or whether they have simply displaced emissions from one location to another. Offset programs must be carefully designed and monitored to ensure that offsets are not approved unless their reductions are permanent, verifiable and additional to those that would have occurred anyway. The next chapter explains how international offset programs can be strengthened to avoid such problems.

Another very important policy choice is the allocation of permits in any cap and trade program. The market value of permits in a national cap and trade program would be between $50 and $150 billion in the program's initial decades. These permit values would greatly exceed actual mitigation costs in the early years. Naturally, how these valuable permits are to be allocated has become a burning political issue. Lobbyists representing various sectors and interests have besieged Congress, whose members have pushed forward the interests of their favored constituents. As a result, in proposed legislation, large fractions of permits would be distributed free to these interests to 'cushion' the economic impacts on influential constituencies, including electric utilities, heavy industry, agriculture, rural electricity users and others.

Such an outcome would be highly inefficient and inequitable. It would transfer valuable assets (permits with a huge market value) onto the balance sheets of these firms, precluding other uses of the potential revenues. Some have argued that awarding free permits is merely a method of distributing benefits and compensating interests that would potentially be harmed by a cap and trade system, without any effect on the system's overall cost-effectiveness. That is only true in a very abstract theoretical model, not in the American economy as it actually exists.

In an upstream cap and trade system, fossil fuel prices would rise to incorporate permit prices whether the permits were given away or auctioned off. If given away, they would still represent an opportunity cost for the recipients, because they could be sold at their market value. That value would have to be recovered if the recipients used the permits directly to sell more fuel. If the fossil fuel sellers received their permits free, it would create a huge windfall profit for the fossil energy industry.[5]

In a cap and trade system focused only on the electric utility sector, the effect on electricity rates would depend largely on whether the utility's sales are in States that have deregulated the industry or in States that have maintained rate of return regulations over electricity rates (Williams, 2008). In the latter, if the utility does not have to pay for its permits or pay more for its purchased power, the public utility commission would not include any permit cost in the utility's rate base, leaving electricity rates unchanged and blunting any incentives that electricity users would have to conserve electricity. If energy efficiency improvements are discouraged, then carbon prices will be higher and the economic impacts greater. If the utility does have to pay for its permits, it would pass along those costs to ratepayers in recovering its fuel costs, encouraging greater attention to end-use efficiency improvements.

In the 15 States that have deregulated their electricity industries, the situation is more complicated. Electricity rates now reflect the marginal cost of meeting demand at all times, so that electricity rates are relatively high during times of peak demand and much lower in off-peak periods. Coal-fired power usually provides steady baseload power and typically represents the marginal supply only in off-peak periods. In those periods, the utility would be able to pass along permit prices fully. In peak periods, rates are more likely determined by the costs of gas-fired 'peaking' plants with fewer emissions per kilowatt-hour generated. Coal-fired plants would only partially recover their permit costs and electricity rates would rise to a lesser extent. If peak demands were met by hydroelectric power, then rates would hardly reflect carbon permit prices at all.

For industrial and other potential recipients, free permits are a lump-sum transfer of assets, leaving the operating incentives of the recipients unchanged. Just as recent infusions of capital into the banking sector left their incentives to lend unchanged, because it did not change the balance of market risk and reward, the free allocation of permits to carbon emitters would not reduce any propensity on their part to raise prices or reduce output in response to higher energy prices. Only the cap on emissions and the resulting carbon price would affect firms' incentives, which would be the same whether or not the firms must pay for their permits. Since free permits would represent a new asset on the firms' balance sheets, the benefit would flow almost entirely to shareholders, not to employees or customers. This is a highly regressive disposition of the asset.

Decisions about allocating permits worth about $1.75 trillion over a decade also have unavoidable macroeconomic implications. That amount of revenue, if captured by the federal treasury through auctions, would make a big dent in the fiscal deficit, with implications for long term interest rates, the dollar exchange rate, inflation rates and other macroeconomic variables.

Repeatedly, studies have found that auctioning permits and using the proceeds to improve the functioning of the economy would substantially lessen any adverse economic impacts of a cap and trade program (Parry and Oates, 2000). Typically, these analyses have examined the possibility of returning revenues to households and firms by lowering the marginal rates of taxes that distort markets. These include taxes on payrolls and wage incomes, which discourage labor supply and demand, or taxes on investment returns, which discourage both savings and investment. The higher the current rates of those taxes and the more differentiated their exemptions and deductions, the greater are their distorting effects and the greater the benefit of using revenues from permit auctions to reduce their marginal rates. Most models represent federal taxes in a highly simplified way, assuming away the myriad exemptions, deductions and special provisions that distort the tax code and encourage inefficient tax-avoiding behavior. These models understate the potential benefits of reducing the most distorting tax rates. Nonetheless, the predicted impacts of doing so are large.

It has also been proposed that permits be auctioned and the revenues be returned to all households equally in a lump sum fashion – the so-called 'cap and dividend' (Boyce and Riddle, 2007). While this is more progressive in its implications than a policy that gives permits to companies, it does not improve the functioning of the economy, because it leaves everybody's incentives unchanged.

LESSONS FROM ECONOMIC ANALYSES

Macroeconomic impacts

In general, even with lump sum distribution of any auction revenues or free distribution of permits, all analyses find that the macroeconomic impacts on gross domestic product, household income and consumption would be small. For example, analyses of the Waxman–Markey bill approved by the House of Representatives find that the reduction in the GDP growth up to 2030 would be only 0.02–0.04 percentage points per year.

This is understandable. As the US economy has become increasingly a producer of services rather than material products, the share of energy expenditures in total expenditures has declined. Therefore, the economy-wide effect of a rise in energy prices would be mild. Moreover, the impact of carbon mitigation policies on delivered energy prices would also be moderate. The energy charge in customers' electricity bill is a surprisingly small share in the total, so the increase in the average total bill would be less than 20 per cent as the carbon cap phases in. Also, the increase in gasoline prices at the pump would be much less than that experienced in recent years, when prices jumped to $4.00 per gallon.

There is also close agreement among models that, relative to the baseline growth rate, the decline in household consumption would be between one and two per cent by 2030 and between two and three per cent by 2050. Therefore household consumption would continue to grow almost as fast as in the reference case, a difference of only 0.01–0.02 per cent per year in the growth rate. It would reach the reference case level after a lag of only one year – in 2031 rather than 2030 (US Environmental Protection Agency, 2009). Analyses that make more favorable assumptions regarding auction revenue recycling and that take into account co-benefits from reduced air pollution and other environmental damages related to fossil fuel use predict that economic growth and consumer welfare might be higher, not lower, if such a cap and trade regime were implemented.

Even these projections of macroeconomic impacts are too pessimistic, in that they don't take into account the damages that would be suffered on account of climate changes if mitigation measures are not taken. They are also too pessimistic in basing projections solely on technological options currently available, with the exception of carbon capture and storage, ignoring any new technologies that may be developed for commercial use in the coming three or four decades.

It's important to realize that in serious economic analysis there is just no support for claims that enacting a comprehensive cap and trade regime designed to lower greenhouse gas emissions by 80 per cent by 2050 would impose excessive or unsustainable costs on the economy or on households.

Impacts on employment

Because labor productivity increases over time, output can be sustained with fewer and fewer workers each year. In the past decade or so, labor productivity in the non-farm business sector rose by an average of 2.5 per cent per year, almost double the rate in the preceding three decades. If non-farm business output remained unchanged, employment would contract by about 2 million jobs per year. Without substantial structural changes in the economy, growth in output and income are needed to support increases in job opportunities.

Therefore, since the macroeconomic impacts on production and income are projected to be small, the macroeconomic effects on employment are also projected to be small.[6] A marginally slower rate of economic growth would tend to have a small adverse effect on the growth of employment. Offsetting that, the increase in energy prices would stimulate a shift from more capital- and energy-intensive industries and production methods toward more labor-intensive industries and production methods. Coal mining, oil and gas extraction and refining, and electricity generation have all become very capital-intensive. Energy-intensive industries, such as iron and steel, cement and chemicals production are also relatively capital-intensive. Industries that would expand in an energy transition, particularly energy efficiency services, construction and solar power generation, are comparatively labor-intensive. Moreover, they have a larger domestic content, stimulating employment in this country.

On balance, a comprehensive cap and trade program with permit auctioning, with revenues recycled through cuts in marginal tax rates on earned incomes, would likely lead to employment gains over time, though the magnitudes are difficult to estimate because of macroeconomic complexities, including changes in international trade. Gains depend mainly on impacts on the economic growth rate, which are expected to be modest, and on the extent of substitution of more labor-intensive for more energy-intensive goods and production processes throughout the economy. A reasonably cautious macroeconomic estimate would be the creation of 3–5 million additional jobs over a period of two decades.

Some factual context about the US labor market will be useful in understanding those estimated effects. There are about 145 million recorded workers employed in the country, about 63 per cent of the potential labor force, but there may be another 10 million undocumented workers as well. Three million new jobs, therefore, would represent only two per cent of employment.

Claims that millions of jobs would be created (or lost) because of climate policies are not meaningful unless a time period is also specified. Though it varies considerably from year to year, employment has grown by about 17 million workers per decade over the past 50 years, but growth has swung from 18 million in the 1980s to as low as 14.7 million in the 1990s to as high as 19 million between 2000 and 2007.

It's important to know, if 3 million new jobs are expected to result from climate policies, whether that would be over 10, 20 or 30 years. If the time span were 30 years, for example, 3 million new jobs would represent only about six per cent of overall employment growth, little more than the current share of energy industries in total employment.

Annual employment growth is volatile over time, because it is the small difference between the much larger numbers of new hires and job separations. During the last decade, between 50 and 60 million new employees were hired each year, but between 50 and 55 million employees left their jobs each year, about half doing so voluntarily. The annual turnover is more than 40 per cent of total employment each year, implying that on average a job in America lasts less than 2.5 years. These numbers raise two important questions. First, do estimates of job creation refer to *gross* job creation, ignoring jobs that might be lost or be left elsewhere in the economy, or do they refer to *net* job creation, the difference between new hires and separations? It makes a big difference. The second question is whether jobs are 'permanent' or are 'temporary'. Although many jobs persist even as new people arrive to perform them, many others disappear each year because of businesses disappearing or restructuring, or because the jobs are inherently temporary.

Some recent studies have predicted a more significant increase in employment, but those estimates are predicated on a 'stimulus package' of expenditures and incentives financed by increased government borrowing (Pollin et al, 2008). Naturally, any such fiscal stimulus will lead to a short term increase in employment. Had the stimulus package adopted by Congress in 2008 been devoted entirely to 'greening' the economy, it might have generated 3–4 million clean energy jobs within a few years. Unfortunately, however, in the stimulus package so far adopted, less than 10 per cent of the total dollar amount was devoted to such energy efficiency and investment programs, compared to 38 per cent in China, 58 per cent in the European Union and 80 per cent in the Republic of Korea (Robins et al, 2009). That was a missed opportunity.

In the longer term, stimulus expenditures have to be financed through higher taxes or result in higher deficits and interest rates, all of which would dampen the employment gains. Nonetheless, to the extent that the stimulus package resulted in cost-saving energy efficiency improvements or in investments with a super-normal rate of return, the result would be higher incomes and a faster rate of economic growth, with positive overall employment effects. Sifting through these complexities, the conclusion is that despite the political temptations, climate policy cannot be either justified or condemned because of its employment effects, which will be marginal. The justification for climate policy has to be the need to stabilize the climate.

Implications for industrial competitiveness

In an upstream cap and trade policy, some industries would be adversely affected, relative to the business as usual scenario, and others would be stimulated. Fossil fuel extraction and refining, and energy-intensive industries would grow more slowly and ultimately decline; renewable energy industries and a wide range of industries that contribute to better use of energy or that are the least energy-intensive would expand more quickly. For coal mining and oil and gas production, under an upstream cap and trade regime a good part of the output decline would be offset by higher prices.

Since energy-intensive industries are involved in international trade and competition, the effects of domestic climate policies on them depend partly on the extent to which higher fuel prices can be offset by efficiency gains and partly on what actions trading partners also take. If trading partners also adopt comparable policies to reduce emissions, competitive effects will be minimal. Should trading partners not adopt climate policies comparable to those implemented in the US, there would be considerable 'leakage' to other countries in the form of shifting trade and production patterns.

108

In some legislative proposals, this potential leakage is combated by allocating permits to energy-intensive industries on the basis of their production levels, creating incentives for those industries to maintain or increase output, or by levying protective tariffs against imports. Since heavy industries, such as metals and chemicals, supply inputs to other 'downstream' industries, such as vehicle manufacturing, these protectionist responses would put their customers at a cost disadvantage to overseas competitors. Higher steel and aluminum costs, for example, are a burden to automobile manufacturers that also face international competition. Of course, protectionist policies also undermine the objective of climate policy to bring about a reduction in carbon emissions and they invite other countries to enact similar provisions.

The US, like other mature industrial economies, has an unfortunate tendency to protect its mature 'sunset' industries, those that are already shifting toward emerging industrial countries. The share of newly industrializing countries in world steel and chemicals production has been increasing for decades, for example, for reasons independent of any climate policies or concerns. At the same time, the steel industry is typically found at the center of trade disputes and demands for trade protection.

Such protection is often at the expense of their technologically advanced 'sunrise' industries, in which wealthy, technology-rich countries like the US usually have comparative advantage. The reason, politically, is that sunset industries and their unionized workers have well-established ties to communities and politicians, but newly emerging industries do not.

The modern theory of international trade points up the dangers inherent in this backward-looking orientation. In industries characterized by falling costs and increasing returns to scale, countries that gain a head start in establishing domestic production may gain a lasting comparative advantage. Domestic production can lead to an accumulation of production skills, the growth of complementary supplying industries, economies of scale, and falling costs created by 'learning by doing' and other learning curve advantages. The cost reductions gained through this head start might lead to a long-lasting and self-reinforcing competitive advantage in world competition: lower costs lead to increased production for export and increased production leads to further cost reductions and further competitive advantage.[7]

This phenomenon is already evident in the nascent wind and solar energy industries. Even though much of the technology was originally developed in the US, China, Denmark, Germany and others have now taken over the lead in production and trade, because they moved more quickly and forcefully to stimulate those industries. In 2009 China produced 40 per cent of global output of solar photovoltaic equipment, 30 per cent of global output of wind turbines and 77 per cent of global output of solar hot water heaters (Renewable Energy Policy Network for the 21st Century, 2010). Unless the US shifts its industrial policy away from supporting sunset industries and provides strong stimuli to emerging sunrise industries, it may be shut out of the world market in these fast-growing sectors. The impact on exports, industrial production and employment would be unfortunate. Many new, high-paying jobs would be sacrificed to other countries.

A comprehensive cap and trade regime would be the foundation for this stimulus to sunrise industries. By creating an assured market space for low carbon energy sources and by raising the costs and prices of competing fossil fuel industries, it would encourage investment in renewable energy industries.

CONCLUSION: HOW TO APPROACH CLIMATE POLICY DECISIONS

The economic analyses that have been carried out on climate policy lead to the strong conclusion that even under pessimistic assumptions about future technology options and adjustment possibilities within the US

110

economy, the economic impacts of policies designed to achieve an 80 per cent reduction in emissions by 2050 will be small and manageable. Economic growth and living standards will continue to improve at almost the same rate, even ignoring the averted damages from climate change and other environmental benefits. Under more optimistic assumptions and taking these benefits into account, living standards will rise more rapidly if effective and economical climate policies are put in place.

There is a common-sense rationale for this important conclusion: the impact of a transition away from fossil fuels is bounded by the share of energy expenditures in the US gross domestic product and the rise in energy costs resulting from a transition to low carbon energy costs. The former is small, well under 10 per cent, and falling over time. The latter is certainly less than 50 per cent. Therefore, the upper bound of potential long run economic impacts is about 4–5 per cent of gross domestic product, even if no adjustments or innovations took place over coming decades. But, as numerous studies have shown, there are many adjustment possibilities, including many that would actually save money in the long run.

The potential damages if this transition is not made are unbounded. If global emissions continue to increase as projected in business as usual scenarios, atmospheric concentrations of carbon dioxide equivalents will reach 800–1000 parts per million by the end of the century. By that time, positive feedbacks, such as melting permafrost, disappearing sea ice, large scale forest fires and altered marine chemistry might make climate change self-perpetuating and irreversible by any available policy options. Though it is impossible to foresee all the possible consequences of a shift in climate beyond those previously experienced in the course of human civilization, the risks are obviously enormous.

The proper framework in which to think about climate policy is a balance of risks and costs.[8] The costs are bounded and relatively small. The risks are unbounded and potentially catastrophic. The costs of an energy transition should be seen as an insurance premium, perhaps costing only one or two per cent of income, against the enormous risks of global climate change. It is a form of social insurance, since individual households cannot by themselves obtain such insurance. When compared to other existing forms of social insurance, such as national security or health insurance, it is obviously affordable.[9]

As most citizens and a growing number of business and political leaders now recognize, the issue is not whether to enact policies to make this transition, but how to design the best policies for the purpose. The analyses underlying this book show that there are better and worse policy decisions. Making better policy decisions can reduce economic costs and impacts by a great deal, by more than half. Enacting a comprehensive cap and trade regime with banking and borrowing of permits, carefully crafted 'offset' provisions, and productivity-enhancing use of revenues from permit auctions is the foundation. Building on this foundation with strategic investments in research and infrastructure and with supportive institutional and policy changes will accelerate the transition and also help to reduce costs.

NOTES

1   Many of these simplifying assumptions are problematic. Many models assume a single 'representative' household, which precludes analyzing how different policies affect the distribution of income across households and consequently spending and employment patterns. In most models, the number of goods and services are severely reduced into a few sectoral aggregates, implying the assumption that various goods and services can be readily substituted for one another at constant prices.

2   See Repetto and Austin (1997) and a later interactive online meta-analysis at www.climate.yale.edu/seeforyourself.

3   Models strictly abstract from climate change damages by assuming that no production costs are ever affected by weather and that households are indifferent to all sorts of weather events.

4   Their most recent analysis can be accessed at http://emf.stanford.edu/research/emf22.

5   One proposal includes a limited carbon tax covering petroleum fuels, with the rate linked to a cap and trade program covering electric utilities and major industrial sources. Such a rate would tend to fluctuate with movements in the permit price.

6   The models have a limited ability to deal with employment issues because they rule out by assumption any unemployment. They assume that all markets are in equilibrium, including the labor market, implying that wage adjustments always balance labor demand and supply.

7   His contribution to this new understanding of international trade helped earn Paul Krugman his Nobel Prize in economics and is now incorporated into standard textbooks. See Krugman and Obstefeld (2004).

8   This is the framework advanced by the influential Stern Review on the Economics of Climate Change, carried out for the British treasury. See Stern (2007).

9   Indeed, the military and intelligence agencies have come to understand climate change as a threat to national security and the medical and public health professions view climate change as a health threat.

REFERENCES

Arthur, W. (2009) *The Nature of Technology: What it is and How it Evolves*, The Free Press, New York

Bollen, J., Guay, B., Jamet, S. and Corfee-Morlot, J. (2009) 'Co-benefits of climate change mitigation policies: Literature review and new results', Economics Department Working Paper 693, OECD, Paris

Boyce, J. and Riddle, M. (2007) *Cap and Dividend: How to Curb Global Warming While Protecting the Incomes of American Families*, Political Economy Research Institute, University of Massachusetts, Amherst, MA

Fawcett, A., Calvin, K. V., de la Chesnaye, F. C., Reilly, J. M. and Weyant, J. P. (2009) 'Overview of EMF 22 US Transition Scenarios', *Energy Economics*, vol 31, ppS198–S211

Krugman, P. and Obstefeld, M. (2004) *International Economics: Theory and Policy*, Pearson/Addison Wesley, Boston, MA

McKinsey & Co (2007) *US Greenhouse Gas Emissions: How Much and at What Cost?*, New York

Paltsev, S., Reilly, J., Jacoby, H. and Morris, J. (2009) *Costs of Climate Policy in the US*, MIT Joint Program on the Science and Policy of Global Change, Report No 173, Cambridge, MA

Parry, I. and Oates, W. (2000) 'Policy analysis in the presence of distorting taxes', *Journal of Policy Analysis and Management*, vol 19, pp603–613

Pierce, B. and Dennen, K. (eds) (2009) *The National Coal Resource Assessment Overview*, US Geological Survey, Washington, DC

Pollin, R., Garrett-Peltier, H., Heintz, J. and Scharber, H. (2008) *Green Recovery: A Program to Create Good Jobs and Start Building a Low-Carbon Economy*, Political Economy Research Institute, University of Massachusetts, Amherst, MA

Renewable Energy Policy Network for the 21st Century (2010) *Global Status Report*, Paris

Repetto, R. and Austin, D. (1997) *The Cost of Climate Protection: A Guide for the Perplexed*, World Resources Institute, Washington, DC

Robins, N., Clover, R. and Singh, C. (2009) *A Climate for Recovery: The Color of Stimulus Goes Green*, HSBC Global Research, London

Schumpeter, J. (1945) *Capitalism, Socialism and Democracy*, Harper Brothers, New York

Stern, N. (2007) *The Economic of Climate Change: The Stern Review*, Cambridge University Press, Cambridge, UK

US Department of Energy, Energy Information Agency (2010) *Annual Energy Outlook*, Washington, DC

US Environmental Protection Agency, Office of Atmospheric Programs (2009) *Analysis of the American Clean Energy and Security Act of 2009*, Washington, DC

von Weizsäcker, E., Hargroves, K., Smith, M., Desha, C. and Stasinopoulos, P. (2009) *Factor Five: Transforming the Global Economy Through 80% Improvements in Resource Productivity*, Earthscan, London

Williams, E. (2008) *Greenhouse Gas Allowance Allocations: Cost Pass-Through, Sector Differentiation and Economic Implications*, Nicholas Institute for Environmental Policy Solutions, Duke University, Chapel Hill, NC

Chapter 5

Next Steps in International Cooperation

In the December 2009 meeting in Copenhagen of Parties to the United Nations Framework Convention on Climate Change, negotiators tried with limited success to establish a road map for the period following the expiration of the Kyoto Protocol in 2012. This chapter explains how an effective agreement can still be found if future negotiations emphasize three main components: first, binding commitments from developed countries to reduce their emissions rapidly; second, commitments by major developing countries to adopt policies and measures that can reduce GHG emissions without hampering their economic growth; and third, reform of the Clean Development Mechanism and other international financing mechanisms.[1]

Many participants and observers at Copenhagen and at follow-up meetings were disappointed that neither a binding international agreement nor any firm commitments to reduce emissions were achieved. The US tentatively offered to make emissions reductions of 17 per cent below 2005 levels by 2020 and further reductions after that date, if Congress enacted legislation to that effect. Other major developed (Annex I) countries, skeptical of America's ability to pass such legislation, made qualified commitments that they would implement only if a comprehensive international agreement is reached in which other developed and developing countries pledged comparable commitments or actions.

Developing (non-Annex I) countries agreed only to submit lists of mitigation actions that they were willing to undertake voluntarily. Brazil, for example, indicated its intention to reduce deforestation significantly and to increase the use of biofuels, hoping to reduce carbon dioxide emissions by about one-third below business as usual projections by 2020. Indonesia also expressed an intention to reduce deforestation and increase energy efficiency. Though monitoring of deforestation has improved substantially through satellite observation, these and other tropical countries have struggled to control deforestation, so these targets are largely aspirational until effective supporting policies and measures are adopted. China indicated an intention, through unspecified actions, to lower its carbon dioxide emissions per unit of GDP by 40–45 per cent by 2020 compared to the 2005 level, to increase the share of non-fossil fuels in primary energy consumption to around 15 per cent by 2020 and to increase forest coverage by 40 million hectares. India indicated an intention to bring about a reduction in its emissions per unit of GDP of 20–25 per cent by 2020 though unspecified voluntary actions.

Both China and India have National Climate Change Action Plans to back up these goals and, as this chapter describes, the Government of China has already enacted many specific measures to try to achieve its targets.[2] Though American politicians often excuse their own unwillingness to pass national climate legislation by pointing to China's refusal to make binding international commitments, China has already adopted more effective policies to reduce emissions than the US has. Some of the most important steps are described in Box 5.1.

Those who were disappointed by the failure at Copenhagen to negotiate binding commitments underestimated the importance of climate legislation in the US, without which the Obama administration could not pledge binding commitments that the Senate might not be willing to ratify, as happened after the Kyoto negotiations. No other Annex I countries would make unconditional commitments without knowing what America would do. Developing countries adhered to

Box 5.1 *What China has already done*

Since 2005, almost all large new power plants constructed in China have used efficient super-critical technology. By 2007, Chinese companies had started constructing even more advanced ultra-super-critical power plants. Since 2007, the Chinese government has closed more than 500 older sub-100MW power plants, each of which emits almost three times the CO_2 per kilowatt hour of the best available technology. China has begun closing relatively inefficient plants with capacities under 300MW, which currently account for 30 per cent of total capacity. The National Development and Reform Commission in China now requires that large power companies 'buy out' and close down inefficient generating plants equivalent to 60 per cent of proposed construction capacity before projects will be approved.

China has completed construction of a next stage generating plant, a 250MW integrated coal gasification combined cycle (IGCC) power plant, with potential for carbon capture and storage. The second and third stages are to add 400MW of IGCC capacity with commercial scale carbon capture and storage.

China's Renewable Energy Law, which came into effect in 2006, mandates that the power grid purchase renewable power. China has already adopted a national renewable energy portfolio standard calling for an increase to 3 per cent of total generation by 2010 and 15 per cent by 2020. China also subsidizes wind, solar and bio-power projects, including a feed-in tariff for wind power.

China aspires to be the market leader in renewable energy. China is a leading manufacturer of photovoltaic (solar) cells, second only to Japan, and became the world's largest manufacturer of wind towers by the end of 2009. China has 60 per cent of global installed capacity of solar water heaters, now installed in 10 per cent of all Chinese homes. The market continues to grow at about 20 per cent per year. In 2009, China invested almost twice as much in renewable energy projects as the US.

In the transportation sector, fuel economy standards were issued in 2005 that were much stricter than those in the US. They have recently been tightened and remain stricter than the new American standards. In 2006 a heavy 'gas guzzler' excise tax based on engine displacement was adopted, the rate ranging from 1 per cent for a fuel-efficient subcompact car to 40 per cent for a large SUV. China is also heavily investing in hybrid electric and all-electric vehicles. A Chinese company is one of the first to introduce an all-electric vehicle to commercial mass markets, both domestic and export. China is also by far the world's largest producer and consumer of electric bicycles.

To reduce energy use in buildings, new constructions are required to use energy-efficient materials and insulation, and to adopt energy-saving technologies for heating, air-conditioning, ventilation and lighting systems. The design standards require new buildings to lower energy consumption per square meter to half the current Chinese average or less, and stricter standards were adopted for richer cities, including Beijing and Shanghai. The national government has also adopted one of the world's most comprehensive mandatory energy efficiency testing and labeling standards for home appliances, as well as energy efficiency standards for lighting, heating and air conditioning and all major appliances.

In 2008 the Government of China adopted a fiscal stimulus package to combat the global economic recession. Nearly 40 per cent of China's $586 billion stimulus plan went toward public investment in renewable energy, low carbon vehicles, high speed rail, an advanced electric grid, efficiency improvements and other pollution controls. This percentage allocation is much larger than that in the comparable US stimulus package.

During the summer of 2010, China's prime minister ordered that more than 2000 relatively inefficient, energy-intensive industrial factories be closed down before the end of September, implying the loss of tens of thousands of jobs. The target plants produced steel, cement and other energy-intensive products. To ensure compliance, these plants would be denied bank loans, export credits, business licenses and, if necessary, electricity (Bradsher, 2010).

their position that without major reductions by Annex I countries, they would not assume any obligations. These basic negotiating stances are not likely to change until the US Congress approves a significant climate bill requiring substantial emission reductions. Although the US government under the Obama administration is willing to join fully in negotiations leading to binding commitments, which could remove a significant obstacle to international agreement, the Congress must enact strong climate legislation in order for other countries to cooperate.

The UN climate negotiations yielded some progress on other fronts, however. Some countries that refused to make national commitments in the Kyoto period have perceived more clearly the risks that climate change will create for their economies. India and China, for example, are increasingly sensitive to the dangers that climate change poses for their limited water resources. Water scarcities already endanger agricultural production and even urban development in some regions. One significant achievement of the Copenhagen negotiations was that these countries participated fully and offered to take significant voluntary steps to reduce emissions.

Countries also all agreed on the need to hold global warming to less than two degrees centigrade to avoid dangerous impacts, implying deep cuts in emissions starting almost immediately. Developed countries agreed to expand international funding to reduce emissions from deforestation and land degradation and to assist the most vulnerable and least developed countries to adapt to climate change. The news from Copenhagen was not all bad.

Nonetheless, major emitting countries, both developed and developing, must go beyond these conditional and voluntary commitments. They must conclude and implement firm agreements to reduce global greenhouse gas emissions sharply. Since the Kyoto Protocol was negotiated more than a decade ago, scientific observations described in a previous chapter have indicated that the processes leading to climate change are even stronger than previously estimated. Climate sensitivity, the response of the global climate to higher greenhouse gas concentrations, seems to be at the upper end of previous estimates. Positive feedbacks accelerating climate change are stronger than scientists thought. As concentrations rise further, the global climate might spiral out of control. The Kyoto Protocol has had little success in reversing the increase in global emissions. Throughout the period, emissions have continued to rise, actually at a faster rate through most of the last decade than during the 1990s. Emissions have risen not only among major countries that did not accept binding commitments in the Kyoto Protocol, including the US, China and India, but also among several countries that did so, including Canada, Spain and Italy. Stronger action is required. In particular, if developing countries' emissions continued to rise as in the past, stabilizing concentrations at 450 parts per million, consistent with the Copenhagen Accord, would be impossible (Edmonds et al, 2007).

OBSTACLES TO AGREEMENT

Despite the urgent need, these continuing negotiations will likely be even more difficult than those that led to the disappointing Kyoto Protocol. In 2010 most of the world was still recovering from a severe economic slump. Financial and budgetary problems in Europe and America threatened to undermine that recovery. Most developed country governments were struggling with budget deficits and increasing indebtedness. Against this backdrop, policymakers have understandably been afraid of placing additional burdens on taxpayers, consumers or producers, in the form of higher energy prices. Many politicians are afraid of losing their positions should they even suggest such proposals. So in the summer of 2010, before the November elections, the US Senate declined even to debate a bill to control greenhouse gas emissions.

Other anxieties that infected the Kyoto Protocol and Copenhagen negotiations are now even stronger. Industrialized countries then feared that their measures to reduce carbon emissions would lead to a flight of industrial investment and jobs to emerging industrial powers in the developing world. In the decade starting in 2001, there has indeed been a significant further shift in manufacturing output and employment. Now China, India, Brazil and other such countries are even stronger industrial competitors. The likelihood has increased that emissions will 'leak' to the developing world through industrial restructuring if Annex I countries reduce emissions disproportionately, without corresponding steps by developing country competitors. Moreover, the very rapid economic growth in the developing world during the last decade has substantially raised its share of global emissions, underlining the futility of emissions reductions only among the Annex I countries.

Therefore, the feasible negotiating space is still extremely constricted. To be successful, negotiations must lead to agreements that countries will voluntarily accept, leading to actions that they will voluntarily implement. Few credible penalties or threats can be aimed at countries that don't fulfill their commitments. Countries that undertake such commitments must judge them to be in their national interests. It is unrealistic to expect that countries will incur significant economic costs in the absence of significant national benefits. Moreover, to be sustainable, agreements must continue to be consistent with national self-interest not only when they are ratified but also in future years, or else countries will desist in their implementation efforts, as several countries have done during the Kyoto Protocol period.

It is now clear that developing countries will not accept commitments that imply significant restrictions on their economic growth. Countries such as Brazil, China and India point with justifiable satisfaction to the rapid growth they have achieved in the recent past and to the rise in living standards that has resulted. In recent decades China has achieved the greatest reduction in poverty in human history. Nonetheless, developing countries point to the large percentages of their populations, more than a billion people in total, still living in extreme poverty and to their need for rapid expansion in employment to absorb these labor surpluses. Countries with economies that have been growing at 8–10 per cent per year justifiably believe that this growth must be supported by increases in energy use, although accepting the possibility that energy use per unit of production can continue to decline, as it has done up to now.

Therefore, developing countries are highly unlikely to accept binding restrictions or national caps on greenhouse gas emissions comparable to those in the Kyoto Protocol for Annex I countries. Those will be seen as almost certain to restrict economic growth, because the base of renewable energy generation is small, and both China and India, among other developing countries, are very dependent on coal for electric power generation. In the event that these countries should prove willing to accept future national emission ceilings, they would insist that those caps would become binding only well into the future. In either case, a strategy that can lead to immediate action is urgently needed.

At the same time, the US and other Annex I countries will not accept that the world's other major emitters continue with 'business as usual' as the underlying principle, as it has been under the Kyoto Protocol. Not only would this risk substantial leakage of emissions to the developing world, it would also imply unacceptably large financial transfers from North to South in compensation for the latter's mitigation costs, whether under the Clean Development Mechanism, a global emission trading system or some other cooperative arrangement. As the US government made clear during the pre-Kyoto negotiations, it will not incur significant mitigation costs unless the major developing countries agree to undertake significant actions as well. The Obama administration has shown no willingness to join in an extension of the Kyoto framework after 2012. If the US refuses to act, other industrialized countries will be limited in their commitments as well.

THE WAY FORWARD

The conflict between these basic negotiating positions leaves only a very limited scope for significant agreement on cooperation during the post-Kyoto period in the decade after 2012. Fortunately, there is an approach that will allow for significant progress in that timeframe despite these constraints. This approach has three main components: first, ambitious binding commitments by Annex I countries to reduce emissions; second, a focus on policies and measures to take advantage of 'win–win' opportunities in developing countries; and third, a reform of the Clean Development Mechanism and other offset programs.

The post-Copenhagen negotiating strategy for the period after 2012 should concentrate on commitments by the non-Annex I countries to implement win–win mitigating policies and measures, in exchange for rapid emission reductions and other measures pledged by Annex I countries. In this framework, developing countries would offer to undertake only those actions in the coming round that would reduce greenhouse gas emissions while promoting, not limiting, their economic growth. Limiting their commitments to these win–win mitigation measures would remove the most significant constraint on their cooperation in stabilizing global concentrations. It could also satisfy demands from Annex I countries that they make significant commitments, because fortunately such win–win opportunities are ample.

By a narrow definition, a win–win measure is one that offers a favorable economic benefit:cost ratio or favorable economic return on investment when considered apart from any value ascribed to the reduction in greenhouse gas emissions it brings about. For example, an energy efficiency investment that offers a two-year payback period, implying a rate of return around 50 per cent, and at the same time reduces electricity consumption and carbon dioxide emissions, is a win–win measure.

A broader definition would extend the economic benefit:cost criterion to include important non-market benefits and costs, such as environmental damages unrelated to greenhouse gas emissions. For example, an investment in rehabilitating a coal-fired power plant to improve its heat rate would reduce carbon dioxide emissions, because it would consume less coal per kilowatt hour produced, but would also reduce other damaging emissions of particulates, sulfur and nitrogen oxides, mercury, and other air-quality pollutants. The resulting improvements in mortality and morbidity, as well as the savings in healthcare costs, would be important economic benefits, even if they were not reflected in the financial costs and returns of the individual power plant. This broader definition of win–win measures extends their potential scope considerably. In China, for example, air pollution, largely from burning coal, causes hundreds of thousands of premature deaths annually. Thirteen of the world's twenty most polluted cities are in China. Air quality in many other large cities, including Delhi, Bangkok and Mexico City, is also life-threatening.

Negotiating on mitigation policies and measures by the non-Annex I countries rather than on targets and timetables for emission reduction is an approach that has been well explored in past years. Although the Kyoto Protocol adopted national targets and timetables, supported by cap and trade programs, as the dominant approach among Annex I countries, a focus on policies and measures emerged as the favored approach at Copenhagen for non-Annex I countries. The International Energy Agency and other bodies have compiled lengthy inventories of policies and measures that have been, or could be, adopted (International Energy Agency, 2010a). Considerable experience and information on the efficacy of many such measures are available.

The many advantages of a focus on policies and measures (PAMs) include the following:

- Commitments on policies and measures are consistent with 'the common but differentiated' responsibilities that countries agreed to in the United Nations Framework Convention on Climate Change.

- They offer flexibility for countries to choose from a wide array of possibilities the measures that make most sense in their particular situations.

- They can be consistent with sectoral approaches favored by some countries, including Japan. For example, countries might pledge to make specific investments or policy and institutional reforms in designated industries.

- Unlike national or sectoral targets and timetables, however, which require further implementing policy decisions, commitments on PAMs specify directly the actions that will be taken.

- For this reason, implementation of commitments on policies and measures is comparatively easy to monitor, because it does not imply tracking an inventory of greenhouse gas emissions.

Negotiations that center on win–win policies and measures can overcome one of the gravest obstacles to successful cooperative agreements on climate stabilization: the temptation for countries to free-ride on the efforts of others. A government that refuses to undertake mitigation measures and incur mitigation costs cannot be excluded from the benefits of other governments' actions to stabilize the global climate. If enough emitting countries decide to free-ride, most other governments will be discouraged from taking action and global emissions will continue to be too high to prevent severe climate risks. However, if negotiations are based on actions that are in countries' economic self-interest, there is little reason for governments to attempt to free-ride.

In return for commitments on policies and measures by non-Annex I countries, the Annex I countries could offer a variety of inducements. These might include:

- More ambitious targets and timetables for their own mitigation trajectories. If fears of leakage of emissions to developing countries are assuaged, Annex I countries could reasonably become more ambitious. Several Annex I country governments have already signaled this intention.

- Less restricted access to their permit markets for Clean Development Mechanism (CDM) projects. The European Union has restricted, and proposed US legislation has limited, the purchase of offset credits from other jurisdictions, partly from reluctance to accept a 'business as usual' baseline that would force Annex I countries to bear the cost of all mitigation actions in non-Annex I countries. This reluctance could be overcome if the latter countries committed to significant policies and measures.

- Greater cooperation on technology sharing and adaptation financing, which are both of high priority to some non-Annex I countries, especially to those with lowest incomes and greatest vulnerabilities.

The approach to negotiations advocated here has an important analogy in trade negotiations under the World Trade Organization (WTO) and the General Agreement on Tariffs and Trade (GATT). Over many negotiating rounds, those negotiations have significantly reduced tariffs and other trade barriers and have contributed significantly to trade expansion and world income growth. With few exceptions, trade negotiations do not deal with targets and timetables for trade expansion; instead they deal with measures, such as tariff and quota reductions, that countries are willing to take in exchange for reciprocal measures by other countries. Monitoring and dispute resolution procedures focus on whether countries implement agreed-on measures and adhere to them. Moreover, trade negotiations are inherently win–win negotiations, because trade barriers injure the economies of those countries that impose them as well as the economies of their trading partners. Countries benefit even when they lower their trade barriers unilaterally, with no reciprocity from other countries, and many countries have done so. In trade negotiations, countries are committing only to do what is in their economic self-interest and that is why these negotiations have succeeded. If climate negotiations adopt the analogous approach, they too can succeed.

THE POTENTIAL FOR WIN–WIN MEASURES

In fact, the potential scope of win–win measures is sufficiently broad that such commitments in the next negotiating round can make a major contribution in both the non-Annex I and Annex I countries. For example, there are abundant cost-effective opportunities for improvements in energy efficiency, using readily available technologies and approaches. The opportunities in the US have been described in an earlier chapter. A broader study for all the G8 countries estimated that the rate of improvement in energy efficiency could be doubled with incremental investments of $400 billion, which would make a substantial contribution to overall mitigation goals. These profitable investments would result in annual energy savings increasing to $500 billion by 2030, with an average internal rate of return on investment of 20 to 30 per cent (United Nations Foundation, 2007).

The potential for cost-saving energy efficiency gains is even larger in important non-Annex I countries such as India and China, where competitive market forces were weaker in past decades and energy prices were controlled. A recent report by the McKinsey Global Institute estimated that developing countries could reduce their growth in energy consumption from 3.4 per cent per year to 1.4 per cent – almost a 60 per cent reduction – just by adopting existing energy-efficient methods and technologies that are cost-saving and pay for themselves. By 2020 their energy use would then be 22 per cent below projected levels (McKinsey & Co, 2008).

These investments in energy efficiency improvements by no means exhaust the potential for win–win greenhouse gas mitigation. In both Annex I and non-Annex I countries, subsidies to energy industries grossly distort demand and supply patterns, leading to inefficient production and excessive energy use. Globally, fossil fuel subsidies increased to $557 billion in 2008 from $342 billion the year before, despite international commitments to cut carbon emissions. According to an International Energy Agency analysis, phasing out these subsidies would result in emissions reductions of 6.9 per cent by 2020, more than the combined emissions of the UK, Germany, France, Italy and Spain (International Energy Agency, 2010b).

The forestry sector also offers other significant win–win opportunities. In countries where forests are already depleted, investment in reforestation is economically attractive in order to provide fuelwood, forage and building materials and to stabilize soils, aside from the benefits of carbon sequestration. In heavily forested countries such as Indonesia, where a large fraction of carbon emissions arise from forest fires and deforestation, forest policy reforms would greatly improve the sustainable returns from forest resources while reducing carbon emissions and increasing sequestration. Illegal logging, overprotection of forest products industries, poor design and supervision of forest harvesting concessions, and conversion of forest lands to unsuitable agricultural uses waste valuable forest resources (Repetto and Gillis, 1988).

Rapid urbanization is bound to continue in the developing world. Two billion additional residents in this century will stimulate enormous growth in urban construction and transportation. Constructing more energy-efficient buildings will amply reward the minor increase in investment costs. The current pattern of urbanization has degraded the quality of city life because of worsening congestion and declining air and water quality. Zoning and land use regulations can promote compact urban growth and thereby promote public transportation and energy-efficient district heating. Investments in buses and light rail, with dedicated bus lanes and prepaid tickets, can be as successful elsewhere as they have been in Latin America.

The conclusion must be that there is ample scope for international negotiations covering the immediate post-Kyoto period on win–win policies and measures that, if adopted, would bring about significant progress between now and 2020 in mitigating greenhouse gas emissions. Focusing negotiations on implementation of win–win opportunities to reduce emissions in non-Annex I countries can be successful in producing agreements that will be implemented and can result in significant reductions. It is one of the few possible approaches with that potential.

In these negotiations, countries could offer specific policies and measures they would be willing to adopt, together with timetables for implementation. Countries could accompany their offers of specific policies and measures with estimates of their probable efficacy in terms of greenhouse gas mitigation. Negotiations would then center on finding an equitable balance of commitments among participating countries. In this approach, a secretariat, such as the International Energy Agency, could provide a negotiating venue; keep track of countries' offers; provide independent resources for estimating the efficacy of particular measures; record and document agreements; monitor and report on implementation; and provide a forum for amending agreements and settling disputes. These functions are analogous to those undertaken by the World Trade Organization in the context of international trade negotiations.

If this approach is adopted for the immediate post-Kyoto period, it is legitimate to ask what the next phase could be after all the 'low-hanging fruit' of cost-saving mitigation measures were harvested. Fortunately, one of the most attractive characteristics of low-hanging fruit is that it grows back. After another decade there will be other attractive investment opportunities and cost-saving measures. Rapid technological progress and innovation, spurred by the flow of investment and entrepreneurial talent into technologies that reduce carbon emissions, will continue to lower mitigation costs, as has happened in past decades. Renewable energy technologies that are now more expensive in financial terms may attain the crossover point of competitiveness with conventional fossil fuel energy sources. There will be other win–win opportunities. By that time, if Annex I countries make good on their post-Kyoto commitments, developing countries may also be willing to accept more far-reaching responsibilities. An opportunity to achieve meaningful cooperation on climate protection now should not be discarded.

CASE STUDY OF POTENTIAL WIN–WIN POLICIES AND MEASURES IN INDIA

Studies in India provide an instructive case study of win–win possibilities, because of India's large and diverse economy. While India is not unique, and other countries have their own range of opportunities, this case study illustrates the extent and range of available measures.

End-use efficiency

Investments in end-use energy efficiency are highly cost-effective in India, because each kilowatt saved is equivalent to almost 1.8 kilowatts of needed generation, taking inefficiencies at the power plant and transmission and distribution losses into account. Deployment of energy-efficient lighting, more efficient refrigerators in households and more efficient motors in industry could alone save as much as 10 per cent of total national power generation.

The overall potential for savings is much larger. According to government estimates, end-use energy efficiency improvements could erase the need for about 23 per cent of generation capacity, with industry and agriculture showing the most potential. A 2001 study found that over a 20-year period end-use energy efficiency gains could profitably reduce electricity demand by 32 per cent and greenhouse gas emissions by 45 per cent, relative to a business as usual scenario. These are win–win opportunities under the narrow or broader definition. For example, initial assessments by the Asian Development Bank found that in the state of Madhya Pradesh energy efficiency can be improved at a cost estimated at 20 to 40 per cent of the cost of the new generating and distribution capacity that would be made unnecessary. Moreover, a significant reduction in life-threatening air pollution from sulfates and particulates would follow (Shukla et al, 2004).

The Government of India has recognized the important potential for end-use energy efficiency savings, partly by creating the Bureau of Energy Efficiency within the Ministry of Power and by enacting the Energy Conservation Act in 2001. The 11th Five-Year Plan (2008–2013) has suggested spending about $1.4 billion for energy-conservation measures over five years, but this is very modest compared to the budget for other elements of the power sector. In addition, the Central Government's Integrated Energy Policy recommends a number of regulatory measures.

For example, in the building sector, construction methods are still typically rather crude and the energy savings potential is enormous (Government of India, 2008b). The Bureau of Energy Efficiency developed and promulgated energy efficiency standards, including a minimum Energy Conservation Building Code, in 2007. The code is mandatory only for new commercial buildings but encompasses envelope, lighting, HVAC, electrical systems and solar devices. In initial case studies, compliance with this code reduces energy use by 50 per cent and raises initial cost by 10–15 per cent with a payback period of 5–7 years, implying a return on investment of 15–20 per cent. Nationwide implementation of this code would obviously be a win–win measure.

According to the Planning Commission, greater end-use efficiency can make a substantial impact not only in buildings and appliances but also in mining, water pumping, irrigation, industrial production processes, haulage, mass transport, lighting and household appliances. Among the concrete policies and measures that would be effective in realizing these gains are the following:

1 Completion of metering programs: installing smart meters on low tension distribution lines and on household and agricultural connections to reduce losses, theft and unpriced electricity use.

2 Completion of tariff reform: eliminating cross-subsidies in electricity tariffs that discriminate against industrial and commercial users and adopting time of use pricing for major consumers. Rate reform would force users to consider actual power supply costs and, in particular, would encourage farmers to raise irrigation efficiency, saving both water and energy.

3 Integrated resource planning: requiring that State Electricity Boards invite bids from distribution companies, energy service companies (ESCOs) and other qualified bidders to supply end-use energy efficiency and to reduce electricity losses on an equal footing with bids to expand generation capacity.

4 Mandatory energy audits: requiring that all large public and private sector consumers carry out energy audits to identify cost-effective energy-saving opportunities.

5 Support for ESCOs and efficiency investments: expanding programs of financial support for energy service companies through credit facilities in industrial banks and payment guarantees for contractual project work with State Electricity Boards and other public sector organizations.

6 Mandatory energy efficiency equipment standards and labeling: phasing in mandatory minimum efficiency standards for important types of equipment, including lighting, fans, motors, pumps and HVAC equipment, which would yield benefit:cost ratios exceeding 5:1, and making energy labeling compulsory for equipment exceeding minimum standards.

7 Mandatory vehicle fuel efficiency standards: gradually increasing mandatory minimum fuel efficiency standards.

8 Mandatory building efficiency standards: directing the Bureau of Energy Efficiency, in cooperation with State governments, to phase in mandatory minimum building efficiency standards for existing buildings exceeding size thresholds, along with penalties and enforcement mechanisms. LEED-like certification programs for energy-efficient buildings could also be promoted.

9 Public sector life cycle procurement: public sector entities use 20 to 25 per cent of all energy in India. Procurement based on minimum life cycle costing rather than the minimum initial cost could be phased in and supported by manuals and training programs.

10 Industrial sector benchmarking and requirements: benchmarking energy-intensive sectors such as metals, fertilizers, cement, chemicals and paper to international standards, and making them eligible for CDM credits for doing better than those standards.

Efficiency improvements in the electric power supply sector

Improving efficiency in power generation, transmission and distribution can realize large financial savings and allow for increased generation and better quality of electricity delivery while reducing carbon dioxide emissions and local pollutants from the power sector. Carbon emissions per kilowatt hour are now 50 per cent higher than the world average. Existing sub-critical pulverized coal power plants are only 29 per cent efficient, and the best 500-megawatt units reach only 33 per cent, well below international standards. Nearly all units could improve their efficiency by 1–2 per cent, which would reduce national carbon dioxide emissions by 3–6 per cent. Similarly, rehabilitating hydroelectric plants could yield much needed peak capacity at a small cost.

Rehabilitation opportunities abound: prevailing energy shortages that make it difficult to take plants off line, the financial difficulties of the State Electricity Boards, the prevalent cost-plus tariff-setting approach, and the shortage of personnel capacity for designing and implementing these projects at the generation plants have all impeded rehabilitation programs. In addition, the primary goals for rehabilitation and modernization projects have been extending plant life and increasing the load factor. Any improvements in heat rate or other efficiency parameters have been incidental. Incorporating energy efficiency among the primary rehabilitation design criteria would result in higher capital expenditures but would improve energy efficiency and would be financially attractive in many cases.

The Government of India has initiated a program to improve operation and maintenance for 10,000 megawatts of poorly performing coal-fired power stations with the lowest plant load factors, followed by rehabilitation and modernization of 5000 megawatts by 2013. The program is definitely win–win, in that capacity could be restored and plant life extended at a cost under $300 per megawatt, much less than the cost of new capacity. A more ambitious rehabilitation program targeting plants with inferior availability and efficiency could restore both at a cost below $500 per megawatt and could reduce carbon dioxide emissions by 25–50 million tons per year.

Retiring the least efficient plants with the lowest availability would also be a win–win measure. The Government of India plans to retire 5000 megawatts of the lowest performing units by 2013. The least efficient 20 per cent of sub-critical coal-fired plants have carbon emissions 60 per cent higher per kwh than the most efficient 20 per cent and an average efficiency almost 40 per cent lower. Because the availability of these plants is so low, they could be replaced by one efficient 300-megawatt unit operating at a high load factor. Doing so would eliminate more than 300,000 metric tons of carbon dioxide emissions (World Bank, 2007a). By 2018, the government proposes to retire or recondition an additional 10,000 megawatts. Even more could be done. The benefits in lower generating costs, increased service reliability and reduced air pollution would justify the added investment, aside from the reduced carbon dioxide emissions.

Poor quality coal is particularly problematic in that it increases operation and maintenance costs while reducing overall efficiency. The use of better quality coals, including washed coals, would pay for itself. Moreover, reducing the ash content of coals at the colliery would reduce rail transport costs that can double or triple the pithead price at longer distances, since high grade energy (electricity) is used to transport low grade coal and ash. The government has enacted requirements for coal washing, but they need to be supported by pricing incentives in the form of sharper price differentials for coal according to caloric value and ash content.

Within the generating sector, however, pricing continues to be based largely on cost-plus formulas that reward inefficiency. The government has attempted to encourage efficiency by using performance benchmarks rather than actual costs for the energy cost component of the tariff. Power plants that outdo the benchmarks make additional profit, because their tariff for the energy cost is based on the norms, not the lower actual costs. Power plants that fall below the benchmarks face revenue loss, because their actual energy costs are higher than the tariff allowed by the regulators. In theory, this system provides strong incentives for power plants to improve efficiency, but in practice it has not worked well and could be made more effective.

For future capacity expansions, super-critical pulverized coal technologies are appropriate for India (World Bank, 2007b). They are commercially viable and there is ample experience worldwide in installing and operating them. Such plants, including scrubbers for sulfur removal and particulate controls, would be at least five per cent more efficient than the best 500-megawatt sub-critical units without flue gas desulfurization. The use of washed coal would increase efficiency by another one per cent.

There are also potential win–win opportunities in making investments in transmission and distribution (T&D), aside from the metering program discussed already, in order to reduce extraordinarily high transmission and distribution losses. Such investments in the period 2002–2007 were only half those for generation, even though many experts advised that investments in T&D be about as large as those in generation. A summary of win–win opportunities in India's power supply sector would include the following:

1 Plant rehabilitation: accelerating the schedule for rehabilitation and modernization of existing plants, with greater attention to potential energy efficiency improvements, and including hydroelectric plants in the accelerated schedule.

2 Plant retirement and replacement: accelerating the schedule for retirement and replacement of the least efficient and least available coal-fired power plants.

3 Capacity additions: establishing and enforcing efficiency standards for all capacity additions.

4 International cooperation: expanding cooperative agreements for imports of hydroelectric power from Nepal, which has large underutilized resources.

5 Transmission investments: increasing investments in transmission and distribution infrastructure to reduce power losses below 10 per cent, with a yield equivalent to at least 10,000 megawatts of new generating capacity.

6 Distribution system investments: accelerating investments in better insulated conductors, capacitators, efficient and low-loss transformers, rationalized distribution networks, and improved metering systems and instrumentation.

7 Expanding wholesale competition: requiring utilities to consider bids from independent power producers as well as energy service companies in adding capacity.

8 Measuring and monitoring performance: requiring all power plants to measure efficiencies routinely, to carry out energy audits to assess their efficiency levels and to submit timely performance data to regulators.

9 Tightening performance incentives: ensuring that the regulatory benchmarks on power plant efficiency are sufficiently stringent and provide strong and consistent incentives for efficiency improvements.

10 Improving coal quality: in addition to existing mandates, providing better incentives for the production and use of higher quality coals by restructuring the coal grading and pricing scheme.

Energy subsidy and price reforms

Energy prices in India are heavily influenced by subsidies and cross-subsidies. Prices do not reflect marginal supply costs and they encourage inefficient energy uses. Energy subsidies have also drifted away from intended target populations and are difficult to justify on equity or other social grounds.

The power sector is the joint responsibility of the Central Government, which controls about two-thirds of generating assets, and State Governments, where State Electricity Boards (SEBs) have oversight authority. The SEBs have been forced by State Governments to pursue political and social objectives, subsidizing power for households and farmers. To attract rural voters, many States introduced flat rate tariffs based on the horsepower of motors or electricity connections rather than on metered usage, so that energy consumption is free at the margin and receipts fall far below supply costs. With these subsidies, the agricultural share of electricity consumption rose from 9 per cent in 1970 to over 20 per cent in 2000.

To try to make good their losses on subsidized power, SEBs set much higher tariffs for large industrial and commercial enterprises. As a result, many firms have installed captive power plants or diesel generators, lowering overall energy efficiency and exacerbating the SEB's financial distress. Average receipts are only about 70 per cent of average supply costs. Consequently, most SEBs have no money to improve power quality or availability and have had to depend on financial bailouts from State and Central Governments to stay afloat. These enormous annual subventions have run at about 1.5 per cent of GDP in recent years.

Subsidies and political interference have also led to corruption. Theft of power and nonpayment of electricity bills, sometimes sustained by bribes to power sector staff, account for about 15 per cent of electricity consumptions. In most States, only about 55 per cent of electricity consumption is billed and only about 40 per cent of the billed amount is actually collected. Theft and pilferage is estimated at about $5 billion annually. Consequently, at least one-third of electricity users, including thieves, delinquents and supposed farmers, face a zero marginal cost for electricity, destroying incentives for efficiency.

The Central Government and some State Governments have been making strong efforts to address these problems, with support from the World Bank and other agencies. Nonetheless, electricity rates are still highly controlled and distorted. The Accelerated Power Development and Reform Program (APDRP) provides incentives and investment support to upgrade transmission and distribution systems in States that enter a memorandum of understanding accepting a time-bound reform agenda. The agenda includes reorganization of the SEB, electricity rate reform, metering of low voltage distribution lines and consumer connections, energy audits, and improved billing procedures. The basic objectives of the program are to reduce distribution losses from the current 30–40 per cent to a more reasonable but still relatively high 15 per cent and to improve supply conditions for consumers. Progress has been made under the APDRP, but much more could be done. It is ironic that Indian companies are world leaders in information technology, but India's crucial power grid is still deficient in its use (Tongia, 2004).

In addition, the electricity supply system lacks the capacity and motivation for demand side management programs. In many States, the utilities are struggling to overcome supply shortfalls to meet growing demands while engaging in the reforms and reorganizations that recent policies demand. Personnel are overwhelmingly supply-oriented with an engineering perspective. Relations with customers are often problematic, due to the poor quality of supply and billing irregularities.

Electricity subsidies are obviously regressive, because about 45 per cent of the population, predominantly the rural poor, still lack electricity connections and receive no benefit from the subsidy. In fact, the resulting financial burden on the SEBs impedes progress in rural electrification. Medium and large farmers gain most from the subsidy for agricultural uses. They are more likely to have tubewell irrigation and to operate pump sets and other electric equipment, while small poor farmers lack connections and may even purchase well water for irrigation from wealthier neighbors.

The rapid growth in electricity and other forms of energy consumption is mainly the result of the improving living standards of middle and higher income households in towns and cities. The disparity in energy use among income groups has widened along with the distribution of income itself. According to a 2007 report, the highest income group is now responsible for carbon emissions of 4.97 tons per capita, only slightly less than the world average of 5.03, and the wealthiest income group's per capita emissions are 3.5 times that of the lower 75 per cent. In international negotiations, the Government of India has maintained that equal per capita carbon emissions represent the only ethically defensible norm for international cooperation, but within India energy use and carbon emissions are becoming more unequal (Ananthapadmanabhan et al, 2007).

Other energy prices are also affected by subsidies (Government of India, 2004). Liquefied petroleum gas has been subsidized as a clean fuel substitute for biomass burning in rural areas, but studies have shown that most of the subsidy actually benefits urban families, while most rural households continue to rely on wood, dung and other biomass fuels for cooking. Subsidized kerosene is used for lighting in rural areas, but is also widely used to adulterate diesel fuel for generators and motors. The benefits go primarily to non-poor households. Targeting of subsidies through household 'Below Poverty Line' cards and certificates has proven ineffective, because such cards are now possessed by a large percentage of households who are above the poverty line.

Phasing out these subsidies would be a win–win opportunity. Inefficient energy uses would be discouraged and a heavy fiscal burden would be relieved, freeing resources for more effective anti-poverty programs. The Government of India announced in 2010 steps to phase down some of these subsidies.

The subsidy on nitrogenous fertilizers such as urea also provides a win–win opportunity for emission reduction, since nitrous oxide is a greenhouse gas 300 times as powerful as carbon dioxide and is generated when urea volatilizes in the field. Fertilizer subsidies were introduced more than 40 years ago to support the Green Revolution introduction of high-yielding fertilizer-responsive seed varieties. Now, however, the use of high-yielding varieties and chemical fertilizers is well established among large and small farmers alike throughout India. The distribution of the fertilizer subsidy among farmers reflects the unequal distribution of land holdings, larger farmers benefiting disproportionately. Moreover, between a third and a half of the fertilizer subsidy doesn't reach farmers at all but protects inefficient segments of the domestic fertilizer industry that cannot meet world prices because of inefficient plants and obsolete technology. Phasing out the subsidy would promote greater efficiency in the industry and among farmers.

Price controls on coal and hydrocarbons present difficult problems of rationalization. Though the principle behind pricing of hydrocarbons has been parity with import prices, the government has been reluctant to allow rising international prices to pass through to consumers.

Adjustments have lagged, sometimes rather badly. This subsidy to the rapidly growing fleet of personal and commercial motor vehicles is partly responsible for increasing congestion on urban roads, extremely poor urban air quality, and shifts away from more energy-efficient bus and rail transport (Government of India, 2006).

The Government of India has also controlled natural gas prices and allocated available supplies among priority sectors, such as petrochemicals, fertilizers and electric power. Pricing has attempted to provide a fair return to producers while keeping costs down for major users and supporting the administered allocation priorities – a difficult task in the absence of market signals and adjustments. The current position is evidently that price decontrol must await a better balance of supply and demand, but that is exactly the balance that price decontrol would bring about.

Pricing and subsidy reform measures have a political cost. Nonetheless, in the context of international negotiations in which other countries might offer comparable subsidy reforms, India might have much to gain and pricing and subsidy reform could be part of the bargain. Win–win reforms in India include:

1 Reform of electricity tariffs: phasing out cross-subsidies, introducing energy charges for all users, competitive bidding at wholesale level, and distribution rates incorporating effective and improving efficiency standards.

2 Phase-out of subsidies for kerosene and LPG: eliminating these subsidies and reallocating the fiscal savings to other, more effective anti-poverty programs.

3 Full import parity pricing of petroleum products: fully passing through import parity prices for petroleum products to final consumers, except for products with minor export markets.

4 Elimination of fertilizer subsidies: phasing out producer and consumer subsidies for nitrogenous fertilizers.

5   Decontrol of natural gas prices: allowing prices to find competitive levels, relaxing administrative allocations, and allowing pass-through of fuel and feedstock cost increases in product prices.

All of the measures described in this case study would actually strengthen and accelerate India's economic development while substantially reducing greenhouse gas emissions. India could reasonably expect assistance from the international community in implementing them, as well as ambitious commitments from other parties.

REFORM OF THE CLEAN DEVELOPMENT MECHANISM

The other main area for agreement on international climate cooperation deals with improvement in the Clean Development Mechanism (CDM)[3] and other 'offset' mechanisms to finance mitigation efforts in non-Annex I countries. Despite its flaws, the CDM has so far provided finance for more than 4000 projects and has stimulated the emergence of project developers, investors, brokers, consultants and market makers who are positioned to provide the human capital and infrastructure for an expansion of these abatement activities, provided the mechanism is improved.

These financial flows will be essential if developing countries are to reduce emissions at the necessary scale and pace. They are economically justified, because in reducing emissions countries are contributing to a global public good – a stable climate – that benefits not only themselves but other countries as well. Because the country making the investment does not capture all the resulting benefits, countries tend to under-invest and even to free-ride on the efforts of others. International cost-sharing can alleviate this tendency. Moreover, the ability of Annex I countries to take advantage of low cost abatement possibilities in other countries greatly lowers the costs of meeting their own commitments and obligations. Economic studies in the US agree that without these international offset possibilities, it would be more costly to reach mitigation targets and domestic carbon prices would have to be much higher.

If negotiations with non-Annex I countries focus on commitments to implement win–win policies and measures, that alone can overcome some of the shortcomings in the CDM. The CDM is essentially a mechanism designed only for project level activities. Despite efforts to develop programmatic CDM activities, such as India's mission to replace all the country's incandescent light bulbs with compact fluorescents, the CDM is not easily extended to policy changes or for broader programs. Yet these are extremely important mitigation options and much more easily covered by direct negotiations on win–win policies and measures.

Moreover, in order to combat the perverse incentives inherent in the CDM (Repetto, 2001), elaborate safeguards have been necessary to ensure 'additionality' and to construct realistic hypothetical baselines describing what might have happened in the absence of the project to be credited.[4] The presumption in the administration of the CDM is that, absent specific barriers, projects that offer an economic rate of return would eventually be implemented without any additional CDM payment. Therefore such projects would not be considered to bring about 'additional' emission reductions.

Under CDM rules, win–win projects and programs are generally ineligible for credit, because if they are financially and economically rewarding without such credits, it is assumed that they would be implemented through normal market processes, unless the proponent can establish that sufficient barriers exist to prevent implementation. This additionality requirement rules out those activities and projects that can reduce emissions at the least cost or the greatest gain, which are precisely the ones that developing countries should first carry out. These would not, unfortunately, be considered additional under CDM rules and would not be eligible for CDM credits. The additionality requirement in the CDM prescribes an economically irrational requirement.

Commitments on policies and measures can remove this irrationality and can be a useful complement to the CDM.

Past efforts to ensure adequate safeguards have raised transaction costs and delayed project approval, creating a large backlog of potential investments, but in the absence of such safeguards, incentives for participants in the CDM to increase 'deal flow' have led to projects of dubious merit. Unlike a normal commercial transaction, neither the seller nor the buyer of emission reduction credits in a CDM transaction has an intrinsic interest in the actual delivery of the 'good', so long as the paper credit is approved. The need for a great deal of external oversight and evaluation limits the expansion of the CDM and makes it difficult to reach the scale needed to accomplish even a stabilization of emissions in non-Annex I countries. These difficulties are absent from commitments on policies and measures, which can be evaluated simply on the basis of their potential efficacy and can be of significant scale, as the India case study illustrated.

EU and US policy is adding to uncertainty about the future role of the CDM. As a result, investments it generated fell by 50 per cent in 2009. There is considerable skepticism in policy circles in Europe and Washington that all CDM projects have resulted in real and additional emission reductions or that the CDM is an adequate vehicle for future cooperation (Council on Foreign Relations, 2008). To pressure major polluters from the developing world to adopt binding emission reductions, EU regulators have threatened to ban the import of CDM credits from large non-Annex I countries into the Union's own Emission Trading Scheme, by far the world's largest cap and trade program and the main market for CDM credits. In Washington, there is reluctance to adopt a system that transfers a great deal of wealth from the struggling US economy to China, India and other rapidly growing developing economies, or to accept 'business as usual' as the baseline from which credits would be estimated.

Nonetheless, despite its limitations, the CDM can contribute to efficiency and equity in global emission reductions. It should be reformed and retained as part of an overall framework for international cooperation (Stern, 2008). Even after win–win opportunities have been captured, there will remain ample low cost opportunities for additional emission reductions in developing countries. Moreover, financial flows under a reformed CDM will provide resources that developing countries can use to finance win–win investment programs or adaptation programs, or both.

Reforming the CDM requires reformulating the emission baseline from which reductions are estimated and countering the perverse incentives in the current system. One way to deal with the baseline problem is to declare certain kinds of projects categorically 'additional' and eligible for credits. For example, solar thermal and solar photovoltaic power generation projects might be considered intrinsically additional, because their generating costs are still higher than those from fossil fuel plants in most situations. Categorical baselines would reduce investor uncertainty and the need for time-consuming efforts to establish that particular projects would not be implemented without a CDM credit. Transaction costs and delays would be reduced, allowing the program to scale up.

Another promising reform would stipulate that developing countries that want to participate in the CDM market would designate the eligible sectors of their economies and establish baseline conditions required for establishments in those sectors. For example, a country might designate its electric power sector as eligible and establish emission caps or emission standards for individual generating plants. Meeting these standards would require an improvement from current conditions. The status quo would not be accepted as baseline. Plants could then obtain CDM credits for further efforts to emission reductions beneath those caps or standards. Sectoral standards or baselines would remove the hypothetical aspect of estimating reductions as well as the 'business as usual' bias in the current CDM. It would also eliminate the need for complicated methodologies to estimate hypothetically what emissions would have been in the absence of the CDM project. The Government of China has announced a plan to bring its electric power sector into a cap and trade system, which would potentially open the door to this CDM reform.

Time is running out. Unless global emissions soon begin to decline, greenhouse gas concentrations will either reach unacceptably high levels or there will have to be drastic and painful emission reductions in future decades. To be successful, negotiations for the immediate post-Kyoto period must result in immediate and significant reductions in emissions from the high income Annex I nations and at least a stabilization of emissions from lower and middle income developing countries. Co-operation is essential. Neither developed nor developing countries can stabilize greenhouse gases unilaterally, no matter how deep the cuts in emissions, if the other bloc takes no further action. Focusing on the negotiating strategies described above can provide a way forward.

NOTES

1   International cooperation to help the least developed countries adapt to climate change impacts and an expansion of cooperative research and technology development are also important. See Global Leadership for Climate Action (2007). These aspects are also well covered in the World Bank (2010); therefore they are not discussed in this chapter.

2   These government reports are useful resources: China National Climate Change Programme 2007, National Development and Reform Commission, Peoples' Republic of China, Beijing; National Action Plan on Climate Change 2008, Prime Minister's Council on Climate Change, Government of India, New Delhi.

3   The CDM allows parties with mitigation obligations in Annex I countries to pay for projects in participating non-Annex I countries that lead to

emissions reduction beyond those that would have been realized without the project and, after those reductions have been verified and certified, to count them toward fulfillment of the purchasers' mitigation obligations.

4　The CDM rules allow only credit for emissions reductions that are additional to those that would have occurred in the absence of the project.

REFERENCES

Ananthapadmanabhan, G., Srinivas, K. and Gobal, V. (2007) 'Hiding behind the poor', report for Greenpeace on Climate Injustice, accessed August 2010 at www.greenpeace.org/raw/content/india/press/reports/hiding-behind-the-poor.pdf

Bradsher, K. (2010) 'In energy crackdown, China to shut 2000 factories', *New York Times*, 10 August

Council on Foreign Relations (2008) *Confronting Climate Change: A Strategy for US Foreign Policy*, New York

Edmonds, J., Clarke, L., Wise, M. and Lurz, J. (2007) *Stabilizing CO₂ Concentrations with Incomplete International Cooperation*, Pacific Northwest National Laboratory, US Department of Energy, accessed August 2010 at www.pnl.gov/main/publications/external/technical_reports/PNNL-16932.pdf

Global Leadership for Climate Action (2007), *Framework for a Post-2012 Agreement on Climate Change,* United Nations Foundation, Washington, DC

Government of India (2004) *Central Government Subsidies in India*, Ministry of Finance, New Delhi

Government of India (2006) *Report of the Committee on Pricing and Taxation of Petroleum Products*, New Delhi

Government of India (2008a) *National Action Plan on Climate Change*, Prime Minister's Council on Climate Change, New Delhi

Government of India (2008b) *Situational Analysis of Commercial Buildings in India*, Bureau of Energy Efficiency, New Delhi

International Energy Agency (2010a) 'Addressing climate change: Policies and measures', accessed June 2010 at www.iea.org/Textbase/pm/index_clim.html

International Energy Agency (2010b) *World Energy Outlook 2010*, IEA, Paris

McKinsey and Co (2008) 'Fueling sustainable development: The energy productivity solution', McKinsey Global Institute, accessed August 2010 at www.mckinsey.com/mgi/publications/fueling_sustainable_development. asp

National Development and Reform Commission (2007) *China National Climate Change Programme*, People's Republic of China, Beijing

Repetto, R. (2001) 'The Clean Development Mechanism: Institutional breakthrough or institutional nightmare?', *Policy Sciences*, vol 34, pp303–327

Repetto, R. and Gillis, M. (eds) (1988) *Public Policies and the Misuse of Forest Resources*, Cambridge University Press, Cambridge, UK

Shukla, P., Rana, A., Garg, A., Kapshes, M. and Nair, R. (2004) *Climate Policy Assessment for India*, Universities Press, Hyderabad, India

Stern, N. (2008) *Key Elements of a Global Deal on Climate Change*, London School of Economics and Political Science, London

Tongia, R. (2004) *What IT Can and Cannot Do for the Power Sector and Distribution in India*, Program on Energy and Sustainable Development, Stanford University, Palo Alto, CA

United Nations Foundation (2007) *Realizing the Potential of Energy Efficiency: Targets, Policies and Measurers for G8 Countries*, Expert Group on Energy Efficiency, Washington, DC

World Bank (2007a) CO_2 *Mitigation Potential of Grid-Supplied Thermal Power Generation Expansion in India*, Washington, DC

World Bank (2007b) *Clean Coal Power Generation Technology Review: Worldwide Experience and Implications for India*, Washington, DC

World Bank (2010) *World Development Report 2010: Development and Climate Change*, Washington, DC

Chapter 6
Winning the Political Battle Over Climate Policy

Why was the US the only Annex I country that refused to ratify the Kyoto Protocol, committing nations to greenhouse gas reductions? Why has the US Congress refused to enact policies that set a price on carbon dioxide emissions, even when the European Union, Japan, Australia, New Zealand, several Canadian provinces and even many of our State governments are doing so? Previous chapters have shown that the obstacle is neither technology nor economics. It is a political failure.

The Congress, especially the Senate, has been unable to overcome short term partisan and parochial considerations in order to solve this critical national and international problem, despite efforts by a few members to provide leadership. The Obama administration supports a comprehensive, market-friendly climate policy like that described in Chapter 3, but energy interests have lobbied fiercely against it and the Senate has not acted. An equitable, effective and efficient policy is possible, but to bring parties to the table, the administration will have to use all its regulatory authority and the other tools at its disposal. This chapter describes the formidable obstacles and the way around them.

The absence of an adequate response to the most serious environmental problem in history is a failure deeply rooted in the broader failure of America's political system. At the root of it all is money. The average Senate election campaign costs upwards of $5 million. That means that to be re-elected your average Senator must raise more than $16,000 per week, each and every week for six years. A campaign for the House of Representatives costs much less, about $1 million, but Representatives face re-election every two years, so the weekly fundraising burden is about the same.

Representatives and Senators say they hate this constant and demeaning fundraising that takes up about a third to a half of their working week. Some elected officials quit politics because they're sick of asking for money, but the Congress refuses to change the campaign finance system even though they have the power to do so. Why don't they? They keep on with it because campaign costs create an almost insuperable barrier to entry by political challengers, who have a much harder time attracting campaign funds. Incumbents typically outspend challengers by about three to one, keeping the incumbency rate at about 85 per cent for Senators and 96 per cent for Representatives. This explains why so many successful challengers are very rich people opting for a career change or well-known athletes or entertainers. It also helps to explain why the current system survives: those potentially worthy candidates who can't stomach the prospect of constant fundraising stay away from political careers, leaving them to politicians with stronger stomachs.

Where does all this money, $16,000 per week, come from? It doesn't and can't come from small donations by ordinary citizens. Small donations, $100 or less, make up only about 15 per cent of campaign funds in House and Senate elections. Most of the rest comes from large individual donations and from political action committees, which are conduits for other large donations from organizations with interests in legislation. These include corporations, trade and professional associations, labor unions, and interest organization such as the AARP, the NRA and the US Chamber of Commerce.

Naturally, politicians claim that their votes are not affected by campaign contributions or by direct lobbying by these interest groups. Interest groups supplying the money usually back them up, at least in public, claiming that their activities are intended merely to provide useful information, to protect the public interest or the consumer or jobs or 'fairness' in legislation. Is this at all credible? Is there no connection between the sources of funds in a politician's war chest and the positions he or she takes on legislation? Would corporate executives trained and selected to maximize profitability spend hundreds of millions of dollars in political contributions with no return?

No, of course not. Studies that have tried to unravel the tangled nexus between money and votes have found that political spending by corporations tends to have the highest rate of return of any potential investment available to companies and organizations with interests affected by congressional action (Alexander et al, 2009). Not only can political support be bought, it can be bought cheaply, for nickels on the dollar, by interests with deep pockets.

So, for example, Senator James Inhofe, Republican of Oklahoma, who sits on the key Senate Committee on Environment and Public Works, received $768,000 in contributions from the oil and gas, electric utility, and mining industries during the 110th Congress, far more than any other member, according to the Center for Responsive Politics. Even though climate change will probably affect rainfall and heat patterns sufficiently to return Oklahoma to the conditions of the Dust Bowl, Inhofe is a persistent and vehement opponent of controls on greenhouse gases, calling climate change a liberal hoax. The second largest recipient on the committee was Senator David Vitter, Republican of Louisiana, who gleaned $528,000 from the same three industries. Although Louisiana dramatically bears the risks of more intense hurricanes and catastrophic oil spills in the Gulf of Mexico, it also derives considerable revenue from oil and gas exploitation. Vitter also opposes federal controls.

The nexus between money and political support is one of the few remaining examples of bipartisanship in Congress. On the important Senate Committee on Energy and Natural Resources, Senators Mary Landrieu of Louisiana and Blanche Lincoln of Arkansas, both

151

Democrats, were the largest recipients of contributions from these three industries. Despite the vulnerabilities in their States, they voted with Republicans to stop the Environmental Protection Agency from regulating greenhouse gas emissions, an authority granted by the Clean Air Act and confirmed by the Supreme Court. The other largest recipient on that committee was Republican Senator Lisa Murkowski from Alaska, an oil-rich State that is actually melting under residents' feet. Senator Murkowski sponsored the resolution that her colleagues on the committee supported to overrule EPA authority.

Three-quarters of the American population believe that Congress is not working in their interests, and with good cause. For example, North Dakota has been called the Saudi Arabia of wind. The wind industry would provide jobs, incomes and royalty payments to workers, farmers and ranchers there, but 93 per cent of its electric power now comes from coal-fired generating stations. As a result, North Dakota has low electricity rates but by far the highest level of sulfur dioxide emissions per capita of any State in the country, twice those of Kentucky, the runner-up for this dubious distinction. These emissions are a cause of lung disease, acid rain and smog that affects ordinary citizens, but North Dakota is also the nation's fourth largest oil producer on a per capita basis. Incumbent industries such as electric utilities, oil and gas, and coal mining have much deeper pockets than ordinary citizens or nascent industries such as wind power, so it is no surprise that the three energy industries donated $343,000 to North Dakota Senator Dorgan during the 111th Congress. Dorgan, a Democrat who also sits on the Energy and Natural Resources Committee, voted for a two-year suspension of the EPA's regulatory authority over greenhouse gas emissions. Other examples abound.

Of course, campaign contributions are just the tip of the iceberg. At a minimum, those contributions gain access for lobbyists to elected officials and their staff members. The number of registered lobbyists in Washington DC has nearly tripled in the past 15 years, from about

10,000 to almost 30,000 now, and spending on lobbyists is doubling every decade, making it one of the fastest growing industries in the country (Repetto, 2007). On climate alone, there were 2780 registered lobbyists by the end of 2009, about five for every member of Congress. Of these, six per cent represented environmental groups, another six per cent represented alternative energy interests and almost all the rest represented industrial interests (Lavelle and Pell, 2009). Many of these lobbyists are former members of Congress, now overtly and lucratively representing the interests that had supported them in office. The mining, electricity, and oil and gas industries together spent more than $300 million on lobbying in 2008 and more than $330 million in 2009.

It doesn't end there, by any means. Energy industries have also bankrolled extensive public relations and disinformation campaigns to weaken public and political support for action on the climate issue. Along with conservative and libertarian foundations, these industries have financed climate deniers, such as Fred Singer, who earlier in his well-rewarded contrarian career also denied a connection between CFCs and ozone depletion and between second-hand smoke and cancer. The stated goal of these public relations campaigns was to convince the public that climate change is just an uncertain theory, not a fact (Pooley, 2010). To this day, these well-financed public relations campaigns continue to grasp at straws to create doubt about an overwhelming body of scientific evidence on climate change.

Industry has also bankrolled right-wing and libertarian 'think tanks', such as The Competitive Enterprise Institute, many of which do little or no research but nonetheless advance scientific and economic arguments against policies to curb carbon emissions. Industry-supported organizations, such as the American Council for Capital Formation, hire economic consulting firms to produce inflated estimates of economic impacts, feeding them worst case assumptions to put into their models.

On most important environmental issues, in the face of impending action opponents have put up a defense in depth. The first line of defense is denial: 'It's not true; the science is flawed or incomplete.' Fighting health regulations, the tobacco industry held this line for decades. If forced to do so, opponents retreat to the second line: 'It may be happening but it's not harmful.' A good example is the memorable statement about carbon dioxide by Fred Smith, the CEO of The Competitive Enterprise Institute: 'They call it pollution; we call it life.' When that fails, the fallback position is that it may be happening but we can't stop it. Climate deniers have long claimed that whatever changes are observed can be attributed almost entirely to natural cycles, not to anthropogenic sources. Over a period of at least three decades, climate scientists working through the IPCC and other groups have built up an enormous body of evidence against this claim. Pushed back further, opponents then shelter behind economic arguments: 'It may be happening and may cause some harm, but trying to stop it would cost far more damage to the economy and is therefore not worth doing.' Chapter 4 has shown how empty and misleading this argument is. Finally, usually only when legislation or regulation seems inevitable, opponents retreat to their bottom line: 'All right, I accept that you're going to do this, but the costs should fall on somebody else, not on me.' On the climate issue, lobbying and public relations campaigns have taken up all these lines, and still do, but the upsurge of lobbying activities around House and Senate climate bills show that industrial opponents are now down to their last line of defense.

These political relationships between industry and Congress are examples of what economists and observers of government call 'rent-seeking behavior', the quest to capture benefits of government actions while shifting their costs onto others. Politicians naturally support and seek to shelter their constituent interests, but if enough of them put parochial interests above national and global interests, then no legislative solution to the larger problem will pass the Congress. An equitable and efficient solution to the climate problem would keep costs down and distribute them proportionately to all energy users.

Rent-seeking behavior is the air that Washingtonians breathe. Corporations use their public affairs budgets to shape government action in their favor. Members of Congress use funds derived from those corporations and other organized interests to keep themselves in office. As freshman Senator Tom Udall of New Mexico admitted, 'people know it in their heart – they know that this place is dominated by special interests' (Packer, 2010). Everybody benefits, except ordinary citizens, workers and taxpayers whose interests are subordinated to those of the big spenders. Ironically, the sources of funds that turn this wheel are those very same workers, consumers and taxpayers, who have almost no say about the use of corporate revenues for lobbying purposes (Repetto, 2007).

Business leaders and members of Congress, most of them intelligent and well-informed men and women, generally recognize the importance of the climate problem. Chief executives from many large and respected companies have publicly acknowledged the need for action to stabilize greenhouse gas concentrations, including a mechanism to set a price on carbon emissions. Nonetheless, they strive to shield themselves and their most valuable constituencies from the costs of reducing emissions. Members from States with large energy industries try to shield their sales and profits. Those from States with mostly coal-generated electricity try to prevent electricity rates from rising. Those from States with clusters of heavy industry try to prevent energy prices from rising more than in other countries, or to block imports from major competitors. Those from farm States try to gain even more federal subsidies than they already have. Unfortunately, the end result of all this rent-seeking is likely to be a policy that is too weak to bring about the needed emissions reduction trajectory and much more costly than necessary in its economic impacts. Equally likely is failure to enact any policy at all.

The coal industry is desperate in its opposition to legislation that would lead to a loss of its only remaining customers in the electric power industry. Coal industry lobbyists have opposed legislation, and, should that fail, have argued for provisions that would keep the price of carbon emissions low enough to discourage utilities from switching to natural gas, which has much lower carbon emissions per BTU. They lobby to phase in any emissions reduction requirements over a long enough period to allow existing coal plants to continue operating. Most of all, however, the coal industry has attempted to persuade legislators and the public that coal can be a clean energy source. The American Council for Clean Coal Energy, an industry creation, is spending more than $50 million per year to make a case for 'clean coal'. It rests on the proposition that carbon emissions can be captured and buried permanently in depleted underground gas and oil fields or other leak-proof geologic formations.

Only a few pilot scale CCS projects are operating anywhere in the world and only a few integrated coal gasification combined cycle plants have been built in the US. Important questions remain about the cost of carbon capture and storage, which might require a substantially higher price of coal-fired power to be viable; about the pace at which this technology can be phased in; and most of all, about the enormous volumes of carbon dioxide that would have to be permanently stored to keep coal-fired power viable while drastically reducing emissions. Not only would that require finding enough safe sites, it would also require building a new network of pipelines across the country and new industrial plants retrofitted onto existing power stations to capture the carbon dioxide. Developing this technology option, if at all possible, is of the highest priority because it will greatly facilitate the energy transition not only in America but also in China, India and other countries now largely dependent of coal-fired power. The coal industry's campaign, supported by legislators from coal mining States such as Wyoming, Kentucky and West Virginia and from States deriving most of their electricity from coal-fired plants, such as Indiana and Missouri, has portrayed the technology as right around the corner, which is certainly not yet the case.

Fortunately, industrial interests are not monolithic on energy and climate issues. In addition to such obvious supporters as the renewable energy industries, segments of the financial sector that stand to profit from trading carbon permits and re-insurance companies exposed to huge payouts following weather disasters, some less likely partners have joined USCAP, a coalition of corporations and environmental groups endorsing cap and trade legislation. Some electric utilities have joined, either because their generating assets are mostly nuclear, like Exelon, or because, like Duke Energy, they've decided that by joining the coalition early they can negotiate a more favorable permit allocation. GE, another coalition member, is an important producer of turbines for wind and gas power.

Despite the obstacles, the House of Representatives did enact a wide-ranging bill in June 2009, by the narrow margin of 219 to 212, overcoming almost total opposition by Republicans. The American Clean Energy and Security (Waxman–Markey) Act enacted measures to reduce emissions, improve energy efficiency, raise the share of renewable emissions, and promote carbon capture and storage and carbon sequestration in soils and forests. The bill would create a price on carbon by setting up a cap and trade system designed to reduce emissions by 17 per cent below 2005 levels by 2020 and by 83 per cent below those levels by 2050. Congressmen Waxman and Markey, with strong support from House Democratic leadership, engineered this close victory with lavish disbursements to various interest groups of the value of the carbon permits that would be created, approximate $1.7 trillion over the program's first decade. Indicative of the intense bargaining involved in the bill's passage is the report that 300 new pages of legislative language were added to the bill at 3.00 a.m. the night before the floor vote in order to secure the last 15 Democratic votes, bringing the total length of the legislation to over 1500 pages (Pooley, 2010, p396). Many of the provisions included to gain the necessary Democratic votes would raise the bill's costs or weaken its effectiveness, or both.

Difficult as this legislative process was, the obstacles in the Senate to passing a comparable bill are far greater. First is the threat of filibuster. At least 60 votes are required for passage, but the Democrats lost that majority with the election of Senator Brown in the 2008 special election in Massachusetts and are destined to lose addition seats in the 2010 midterm elections. Despite previous support for climate legislation by Senators McCain, Warner and Graham, the Republican Senate leadership seems intent on denying the Obama administration any major legislative victories, hoping thereby to retake majority control in Congress and eventually the White House as well.

The decline of bipartisanship in Congress does not mirror greater polarization among the voters, most of whom describe themselves as moderate or independent and are frustrated by gridlock in government. The decline has its own dynamic. Especially in solid 'Red' or 'Blue' States, low turnout by frustrated voters, particularly in primary elections, leaves disproportionate influence in the hands of small numbers of party activists: 'the base'. In those solid States, primary elections are often more determinative than general elections of the outcomes, so politicians are pushed toward the agenda of the base to ward off more extreme challengers, not toward the political center, and candidates more acceptable to the base are likely to prevail, as happened in the 2010 primary elections for Republican Party candidates.

The decline of Congressional bipartisanship also feeds on itself because legislators in a minority party have diminishing chances of a significant role in shaping proposed laws, in gathering campaign contributions and in exercising other governing powers. Following an election in which one or the other party loses its majority, quite a few legislators decide that the game is not worth the candle and announce their decisions not to seek re-election. For those that remain, regaining majority status becomes all-important, tempting them to adopt the tactics of demonizing, denying and destroying the opposition – whatever it takes to win in the next election. In this political climate, the public interest can easily fall by the wayside.

Quite aside from the partisan split between the two parties, the prospects for climate legislation in the Senate are strongly affected by its distribution of votes, which differs drastically from that in the House of Representatives. It was remarkable that 25 per cent of the votes for the Waxman–Markey bill came from just two States – California and New York. Those two States have only four per cent of the votes in the Senate, no more than those of the two Dakotas. Compared to the distribution of population, the coastal States are grossly under-represented in the Senate and the interior States are over-represented. Those interior States tend to be more resource-dependent. On the one hand, this makes them potentially more vulnerable to the damages from climate change. On the other hand, however, it also makes them more sensitive to changes in energy prices, markets and availabilities.

Many of these interior States were narrowly in Obama's victory column in 2008, including Illinois, Indiana, Ohio, Iowa, Michigan, Wisconsin, Colorado, New Mexico and Nevada. Many of these are swing States that the administration would need to win again in 2012, and as a group they are considerably more dependent on coal-fired electricity and heavy industry than those along the coasts. Gaining the support of their Senators, even if Democratic, would be a tricky political problem for sponsors of climate legislation and would require costly concessions to resolve. For example, 'Blue Dog' Democratic Senators from Michigan, Ohio, Wisconsin and some other interior States wrote to the President stating conditions for their support of climate legislation, including binding emission limits in all major emitting countries, including China and India, and trade sanctions against countries not meeting those limits. These conditions are potential show stoppers in international negotiations.

Enacting a truly transformative climate law, one that sets the US firmly on a steep trajectory of emissions reduction, would be uncharacteristic of the American political system. Such breakthroughs are extremely rare and occur only in a special set of circumstances, none of which are apparent today (Repetto, 2006). First, there must be room on the political agenda. Now, however, the country is in the grip of a major recession, with unemployment and mortgage foreclosures still high, the domestic and international financial system apparently still at risk, and economic issues high on the list of voter concerns. Financial reform, healthcare reform and economic stimulus legislation have been in the forefront. Simultaneously, the country is fighting two Asian wars with extensive daily media coverage. In 2010 politicians were preparing for midterm elections and in 2011 potential candidates will begin positioning themselves for the next presidential election. On top of all that, fortuitous headline events occur, such as the Deepwater Horizon oil spill, which might have provided the Obama administration with ammunition for a clean energy bill had he not handicapped himself by coming out just months earlier with support for expanded offshore oil drilling. There has been little room on the political agenda for a far-reaching climate bill, which explains the decision in 2010 to sideline such legislation.

Second, there must be strong public demand for action. Since it is much easier in the Senate to block than to pass legislation and because legislators are wary of casting votes that will leave them open to attack, important laws are almost always passed in the wake of a strong upsurge of public demand, amplified by a cascade of media attention. Under these conditions, many legislators and interest groups may decide to get onto the bandwagon before it rolls over them. Currently, these conditions don't exist; this upsurge of public demand is missing. Most people are more concerned with jobs, health and economic security than with environmental protection, and even among environmental issues, climate change does not head the list of popular concerns.

In polls, about 75 per cent of Americans express the opinion that the government should take action to stop climate change even if it involves some economic cost, and 77 per cent think carbon dioxide should be regulated as a pollutant, compared to just 8 per cent who favor no action (Leiserowitz et al, 2010). Nonetheless, most of those who favor government action would not be in favor of higher energy prices or a gas tax. There is also a wide spectrum of beliefs about climate change, ranging from alarmed to dismissive in about equal proportions. A majority of those who are dismissive of the issue are Republicans and identify themselves as evangelical or born again. Moreover, recent trends have been in the wrong direction: since the onset of the economic recession, the percentage of respondents who are dismissive of global warming has almost doubled, while the percentage of respondents who are alarmed has fallen by almost half. Even more alarmingly, the percentage of people in America who think that climate change is due to natural causes is as high as the percentages of people with similar opinions in India or Nigeria.

At this point, the American public generally has a relatively shallow understanding of the climate issue and a broad but low level of concern. A majority of Americans feel that more information about climate change is needed. There is widespread confusion between weather and climate, leading people to take a winter with unusually heavy snowfall as evidence against climate change. An important underlying reason for people's lack of concern is that most people think that damages and risks from climate change will appear in the future, perhaps decades into the future. As Chapter 7 discusses, people find it hard to draw a connection between extreme weather events happening right now, such as floods, heatwaves and intense storms, and climate change that is altering the frequencies of these damaging phenomena. The most obvious connections, such as polar bears on disappearing ice flows, seem remote and relatively inconsequential. People naturally subordinate such concerns to more pressing immediate issues and so respond to public relations campaigns financed by energy interests threatening job losses and higher energy prices.

Media coverage of the climate issue has not amplified public concerns but has served to diffract them. Cable news and the internet have fragmented and proliferated available information sources. Many such sources have political agendas, biases or leanings, and most people choose sources that reinforce pre-existing opinions. Moreover, as media revenues shrink, newsrooms are pared back and fewer journalists remain who can write with any depth or perspective on climate issues. Too often what results are lazy stories that strive for a pseudo-balance by contrasting one statement on climate change, perhaps backed by considerable scientific or economic research, with a contradictory one solicited from a well-known opponent of climate action. The journalist behind the story can lay claim to balanced reportage without having to sort through the issue to arrive at an accurate portrayal of what seems to be correct. Moreover, controversy, even if contrived, makes a more interesting story than consensus any day.

That assumes that the journalist can get the story past editorial gate-keepers at all. Climate change is an ongoing, long term problem. Editors characteristically ask what the 'news' is and what is the 'hook' strong enough to compete with other breaking news for airtime or a place in the A-section of the next day's newspaper. Some long term climate reporters claim that when they pitch a story to their editor they get a 'MEGO' (My Eyes Glaze Over) effect as he turns the story down. There is a Catch-22 dynamic in media coverage: in the absence of strong public interest, there is little incentive to cover the issue, but without media coverage or some dramatic precipitating event, public interest is unlikely to build.

Despite the committed efforts of some leaders in the Congress and in the Obama administration, it does not seem that the US is approaching a tipping point at which strong public demands or a national consensus will push legislators, with their own and constituent issues to protect, to enact a strong climate policy that mandates sharp reductions in emissions and creates a meaningful price on carbon. Lacking this momentum, some politicians would advise against expending much political capital on the issue. More typical of political action in such conditions is an incremental response, like the energy bills passed in 2005 and 2007. Such measures might include additional incentives for renewable and nuclear energy, government spending on research and needed infrastructure for the energy transition, heightened standards and incentives for energy efficiency improvements, and the like. Incremental responses implying modest changes to the status quo are by far the most common legislative approaches to just about all issues. Dramatic policy changes are rare.

In these circumstances, more ambitious legislation creating a national cap and trade system that might be enacted into law would probably have so many qualifications and carve-outs that its real effectiveness would be far less than would be apparent. For example, the emissions-reduction trajectory might be back-loaded to future decades; there might be provisions to keep retail electricity prices from rising, thereby weakening the incentives for end-use energy efficiency; there might be permit allocations that shield existing coal-fired power plants and energy-intensive industries; there might be a low ceiling on the allowable price of carbon, weakening incentives across the board; there might be liberal use of emissions offsets, including those that reward activities that would take place anyway. Many of these qualifications found their way into the Waxman–Markey bill as concessions to interest group pressures. Faced with strong and conflicting pressures, legislators often seek compromises that headline the appearance of action but blunt its impact with provisions buried deep within the bill's text.

Legislation is the art of compromise, many would respond. The political process is one that accommodates various conflicting interests. True enough, but faced with the escalating climate problem, a weak and incremental response is not enough. Slowing the growth of emissions is inadequate. Emissions must begin to decline soon and continue to decline at a rapid rate for decades. This implies a complete departure from business as usual. The transportation fleet must move rapidly toward hybrid and electric power trains. Solar, wind and other renewable energy sources must rapidly penetrate the electricity grid and off-grid uses. The entire economy must ramp up the pace of energy efficiency gains. Making this happen demands strong policies, not incremental change.

THE WAY FORWARD

Public demand for action

Strong policies will not emerge unless the public demands them urgently enough to overcome the forces of resistance. Experts in political messaging and issue framing advise against momentum-building campaigns about the dangers of climate change, which their polls find do not resonate with the public. Instead, they advise messages framed around energy independence, clean energy and the jobs to be had in the new energy economy. This advice is almost surely wrong.

Every president since Richard Nixon has warned about the growing dependence on imported oil, with no impact whatever as the share of imports in oil supply has steadily risen. Most Americans don't really care where the gas that fills their tank comes from as long as its price is steady and not too high. The 2008 presidential election campaign demonstrated how easily campaigns for energy independence can be diverted to sloganeering about 'Drill, baby, drill!' or by the observation that coal is America's abundant domestic energy supply.

The Deepwater Horizon oil spill dramatically revealed the environmental hazards of oil drilling in deep waters and has led to some regulatory reforms and rethinking about such drilling off the Atlantic and Pacific coasts. Shutting down the thousands of wells already pumping oil out of the Gulf has not been proposed, though, and even a temporary moratorium on new drilling encountered stiff political and judicial resistance. Across the country most voters believe that environmental quality where they live is generally pretty good, and they are right. The environmental justice movement has proven that where environmental quality is pretty bad are places where low income and minority households live. Unfortunately, those households are less likely to vote or to participate in public debate.

A clean energy message is unlikely to carry the day. Even clean energy installations generate environmental concerns. Voters in Vermont dislike the idea of wind towers on their hilltops as much as well-off vacationers on Cape Cod dislike seeing them on the horizon offshore. Biomass generating plants run into local opposition from people worried about particulates and toxins in the exhaust gases. Even an effort to locate a solar thermal generating facility in the Mojave Desert ran into environmental protests from wilderness and biodiversity advocates. It must be acknowledged that all large scale energy systems have environmental impacts, but some have much more serious impacts than others.

Framing a message around the jobs to be created by an energy transition has a lot of appeal, especially when the unemployment rate is nine or ten per cent, and – as Henry Kissinger might have said – it has the additional advantage of being true. The labor content of industries improving energy efficiency or generating renewable power is greater than that of the fossil fuel industries they replace. Moreover, the import content of the former industries is lower, especially in the construction and installation stages of the work. However, as with most economic arguments, there is another side to that coin. Jobs would disappear in the fossil fuel industries and even a marginally lower rate of economic growth, should that be the result, would imply a marginally slower rate of employment growth. In any case, since the average citizen has difficulty disentangling these economic arguments, it is easy for opponents, even without any factual justification, to label a cap and trade regime or any other limits on carbon emissions as a 'job killer'.

The challenge of building public support for strong climate policies must be faced directly. Campaigns like those mounted by the Alliance for Climate Protection and many other public interest groups must continue and intensify. People must be convinced of the connection between the increasing frequency of damaging weather events that the country is already experiencing – the intense thunderstorms, the floods, the hurricanes, the heatwaves – and climate change. People must be convinced of the consequences of a continued build-up of atmospheric greenhouse gases and understand that current trends are unsustainable and unacceptable. Political messaging can't dance around that central concern. This is by far the greatest environmental challenge in American history and the Obama administration should use its 'bully pulpit' courageously to say so. When the public understands that, it will demand action.

The case against coal

Despite an expanded educational and messaging campaign, building public understanding will take time and there is no time to spare. What can be done now with the means at hand? The key to the solution is coal. Higher fuel efficiency standards and the continuing improvements in the technology will push vehicle sales increasingly toward plug-in hybrids and electric cars, but electrifying the transportation fleet will not help enough if that electricity is generated with coal. Rapid emission reductions will be impossible if new coal-fired plants are built without carbon capture and storage, or if the existing fleet of inefficient coal-fired plants continue to spew out carbon dioxide. Coal-fired power must be the highest priority immediate policy target.

An equitable policy package can be put together for coal interests. An upstream cap and trade system involving permits for coal and other fossil fuel sales would, in effect, create a government-enforced cartel for the coal industry, enabling it to offset lower sales with higher prices and profit margins. Beyond that, if legislators grant coal companies even five per cent of permits free of charge, rather than allocating all permits by auction, short term profits in the coal industry could be maintained while sales declined. In addition, the policy package should strongly encourage carbon capture and storage, the only technology offering the industry hope of longer term viability in a carbon-constrained world. The Federal Government should finance basic research into the geology and geochemistry of carbon sequestration in order to resolve the issue of safe long term sequestration of high volumes of carbon dioxide. It should also offer significant incentives to coal-fired power plants that install and operate carbon capture and storage systems, including resolution of long term liability issues.

Most of these policy inducements have already been put on the table without lessening opposition from coal interests and their representatives. Those interests have demonstrated a determination to resist relentlessly, using whatever means necessary to fight against a strong national climate policy. This is a fight that the administration and the nation cannot afford to lose. In order to bring coal and its adherents in Congress to support a comprehensive climate bill, there must be sticks as well as carrots. Fortunately, the executive branch has assets that it can bring to such a fight.

The most powerful asset is the possibility of imposing regulations on greenhouse gas emissions. The Environmental Protection Agency is now drafting these regulations, which will apply to major emission sources, particularly large coal-fired power plants. From past experience, the electricity industry knows that such command and control regulations, with little flexibility and no possibility of offsetting emissions or acquiring valuable permits, would be much more expensive for them than a cap and trade regime. Naturally, if such regulations are issued, they will be tied up in court for years, but ultimately utilities would be forced to comply. Their leaders know that cap and trade regulations and the policy package described above would be a more favorable outcome. That is why the industry's friends in Congress have tried –

unsuccessfully – to keep the EPA from regulating greenhouse gases. The administration can and should use the possibility of regulation strategically to bring coal interest into agreement.

In addition, from the mine to the power plant smokestack, the coal industry has other significant environmental impacts that are already subject to federal regulation. In the past, such regulation has usually been excessively lax, and many of the environmental damages the industry generates have been off-loaded onto the public. As a result, the true costs of coal-fired power have not been reflected in market prices. A recent National Academy study found that, even excluding climate change, if the environmental costs of coal-fired power were incorporated into electricity prices, rates would rise by more than $0.03 per kilowatt hour, making wind fully competitive (National Research Council, 2009). In order to make coal-fired power less attractive relative to more environmentally benign alternatives, these regulations and their enforcement should be tightened. The executive branch already has the statutory authority to do so.

Worker safety in coal mines is regulated by the federal Mine Safety and Health Administration. After the Sago mine disaster in 2006, which killed 13 miners, it came to light that in the previous year the mine had been cited for 208 safety violations, almost half of them serious and substantial, without paying significant penalties or having to shut down. During the Bush administration, the MSHA had been staffed with many industry insiders and exhibited 'regulatory capture': excessive deference to the coal mining industry. In the wake of the disaster, legislative amendments increased the civil and criminal penalties for mine safety violations and the use of shutdowns in enforcement actions. Within the limits of existing law, the Obama administration should markedly step up inspection and enforcement activities.

In Eastern coal mining regions, especially in Appalachia, the predominant method is now mountain top removal, in which huge earth-moving equipment and blasting remove massive amounts of dirt and rock to expose underlying coal seams. The dirt, rock and mining residues are pushed into adjacent valleys. Over 1200 miles of the region's streams have been destroyed. Rain leaches toxic and acidic waters downstream from these wastes, polluting household water sources and creating significant health problems for residents. For example, residents of coal mining communities in West Virginia were found by public health researchers to have a 70 per cent increased risk of kidney disease and other elevated health risks as well, even after controlling for smoking and socioeconomic status (Science Daily, 2008). Slurries of coal washing wastes are also impounded in these valleys, and they too are subject to leaching and leakage, polluting streams and drinking water supplies. This practice is regulated under the federal Clean Water Act, but obviously not with sufficient rigor to control or reform the process. In April 2010, the Environmental Protection Agency issued new guidance on permitting the disposal of mining effluents and wastes into water bodies. Permitting requirements and enforcement of regulations can and should be tightened.

Most American coal is now strip-mined, creating issues of land reclamation. The federal Surface Mining Control and Reclamation Act of 1977 creates authority, now shared with State governments, to set reclamation standards and require miners to submit reclamation plans with financial assurances for their implementation. A recent report by the Government Accountability Office found that the use of this authority in Appalachia has not always been adequate to prevent long term impacts on water quality and to ensure restoration (Government Accountability Office, 2010). The study also identified other regulatory provisions that can be brought to bear to help achieve these objectives.

Coal-fired power plants are already extensively regulated under the federal Clean Air Act for environmental impacts other than carbon dioxide emissions. Tightening of these regulatory programs would be intrinsically beneficial and would simultaneously encourage utilities to consider other generating options. The Clean Air Act protects against visibility impairment in 156 national parks and wilderness areas located from Massachusetts to California. Fine particles in power plant exhaust can markedly lower visibility in such pristine areas, even over a radius exceeding one hundred miles. The National Park Service has authority to approve or disapprove proposed new generating plants. Existing plants are required to install the best available retrofit technologies to reduce impacts. Strict enforcement of these requirements would have a considerable impact on the siting of new power plants over much of the country.

Several other provisions in the Clean Air Act can also be brought to bear. The Prevention of Significant Deterioration program was adopted to prevent air quality in pristine areas that meet national air quality standards from deteriorating because of an influx of large new polluting sources. Proposed new sources or major modifications that might affect air quality beyond some significance level must carry out a detailed atmospheric modeling analysis to assess potential impacts and install the best available control technology. In September 2009, the EPA proposed that this program would be extended to greenhouse gas emissions and would apply to new or expanded emission sources with annual emissions exceeding 25,000 tons per year, which would include coal-fired power plants. A stringent definition of best available control technology would tend to discourage such investments, especially if accompanied by aggressive standards to control the conventional air pollutants.

A related Clean Air Act program requires that new plants or major modifications that would increase emissions significantly undergo a New Source Review and conform to New Source Performance Standards. This program originated when amendments to the Clean Air Act were passed in 1977. Lawmakers recognized that when earlier legislation limited strict emission control requirements to new plants, a perverse incentive arose to keep dirtier older plants in operation longer. The assumption had been that such plants would be phased out and retired, but experience showed – and still shows – that avoiding pollution control costs provides older plants with a sufficient cost advantage to offset their inefficiency. As of the end of 2009, 45 of the 50 power plants with the most carbon dioxide emissions were more than 30 years old; 5 were commissioned before 1960, another 12 before 1970, another 21 before 1977, and 6 more between 1977 and 1980 (Abrams, 2009).

The utility industry has tried to maintain this advantage by claiming that major modifications to extend the life or production of these plants were only routine maintenance, exempt from New Source Review. The EPA finally sued at the end of the Clinton administration, but the Bush administration chose not to pursue the suits and gutted the regulation, prompting lawsuits from several States downwind of these polluting power plants. The EPA now plans to revive New Source Review and extend it to carbon dioxide emissions, not just particulates, sulfur and nitrogen oxides. Doing so will put pressure on utilities to shut down and replace these dirty old plants.

In addition, coal-fired power plants are the largest source of mercury emissions in the country, as well as emitting other toxic heavy metal substances such as lead, arsenic and chromium. Since mercury diffuses in the food chain, exposure to this neurotoxin is widespread, even to infants in the womb. Under the Clean Air Act, the EPA is required to regulate power plant emissions of mercury and other toxic substances, but has not yet done so. The law requires the EPA to define, and power plants to use, the maximum available control technology, regardless of cost. This requirement could result in a 90 per cent reduction in mercury emissions. The Bush administration attempted to weaken these requirements but was rebuffed in the courts. The EPA's current schedule is to issue mercury regulations in 2011, and to accelerate regulation of other air toxins. This should be done without delay.

Forty years after passage of the Clean Air Act, substantial parts of the country still do not meet the health-based National Ambient Air Quality Standards, either for ozone or for the fine particulates that lodge deep in the lungs and cause the worst health problems. Much of

California and the Eastern seaboard from Virginia to New Hampshire are non-attainment areas, as are other urban clusters around the country. Coal-fired power plants are major contributors to both problems. Along the East Coast, much of the ozone non-attainment problem emanates from dirty coal-fired power plants upwind to the south and west, emitting nitrogen and sulfur oxides that turn into particulates and smog. The Clean Air Act requires States with non-attainment areas to revise their implementation plans to demonstrate how they will come into compliance but, since these areas have persisted for decades, the process clearly does not work, especially not with respect to pollutants that drift across State boundaries. Existing pollution sources in non-attainment areas must install reasonably available control technologies. Proposed new sources must meet the lowest achievable emissions rate and also offset any residual emissions by reductions from other sources. State governments resist adopting these measures out of concerns about economic growth, but it is scandalous that 40 years after the passage of the Clean Air Act citizens are still suffering health consequences from polluted air. The EPA should bear down on these non-attainment areas and on the coal-fired power plants responsible.

In the summer of 2010, the EPA took a giant step in this direction by proposing a new Clean Air Interstate Rule that would force coal-burning power plants upwind of non-attainment areas to install scrubbers to reduce sulfur emissions and catalytic reduction equipment to prevent nitrogen oxide formation, the pollutants responsible for fine particulates and smog downwind. These proposed rules are stricter than those proposed by the preceding Bush administration, which were struck down by the courts. The compliance costs to the industry are reckoned in the billions of dollars, but the rules are estimated to prevent 14,000 to 36,000 premature deaths and greatly reduce nonfatal respiratory illnesses (Broder, 2010). Interestingly, electric utilities successfully advocated a cap and trade approach to this regulation, though opposing such a system for carbon emissions. It is likely that some older, less efficient coal-fired plants will be shut down in response to this rule, when utilities weigh the benefits of the additional investments needed for compliance against the possibility of future carbon dioxide regulations. Regulatory uncertainty can be a powerful deterrent to billion dollar investments, especially when safer and cleaner alternatives are available.

The executive branch can also enlist the financial community more successfully in this effort by vigorously enforcing the 2010 Interpretive Release by the Securities and Exchange Commission, which reminded companies that they are legally obliged to disclose to investors all known material risks to their business from climate change (Securities and Exchange Commission, 2010). These risks might include financial impacts of future regulations, effects on their production costs or product demands, and physical risks to their facilities and operations from extreme weather. In the past, companies and the SEC have been lax about obligatory environmental disclosures and, as a result, most investors have regarded environmental risks as immaterial (Repetto and Austin, 2000), even when they are not. If electric utilities with coal-fired fleets have to internalize all the environmental costs described above, their operating profits would be greatly affected (Repetto and Dias, 2006). The growth of the Investor Network on Climate Change and other investor activist organizations has put more pressure on companies to disclose, explain and manage their environmental exposures. The SEC can support this trend through active enforcement of its own rules.

Environmental and consumer activists at State and local level should contest every permitting process for coal-fired power plants, both proposed new plants and expansions or re-permitting of existing plants. The EPA's finding that greenhouse gases endanger public health and welfare and are subject to regulation under the Clean Air Act provides activists with solid grounds for opposition: environmental grounds, that coal-fired plants create an environmental hazard, and economic grounds, that coal-fired plants will be uneconomic when and if regulations are issued, leaving ratepayers with stranded costs. Not all opposition will succeed, of course, but even unsuccessful attempts will cause delays, increase costs and raise questions in the minds of both Public Utility Commissioners and electric utility executives as to whether coal is the most prudent generating choice when billion dollar investments are on the line and natural gas or wind is increasingly available. This can succeed. Forty years ago similar activism nearly shut down the nuclear power industry in the country.

There is a way forward to overcome the political obstacles to an effective climate stabilization policy, but it runs uphill. It involves the recurring battle between the diffuse public interest and the organized interests. Few readers of this book will be deeply involved in the regulatory issues described above, but they should nonetheless wonder whose interests are being represented by the thousands of lobbyists working on climate and energy issues in the nation's capital and what return is expected from the hundreds of millions of dollars of political spending by energy industries. They should be concerned to know whether their elected representatives are actually representing their interests and those of their children and grandchildren or the interests of major campaign contributors. They should question why large corporations that don't hesitate a minute before raising their own prices or laying off their own workers are funding political campaigns against climate legislation in the name of job protection or price stability. The way to resolve the climate problem goes uphill, but it will go faster the more people get behind the solution and push.

REFERENCES

Abrams, C. (2009) *America's Biggest Polluters: Carbon Dioxide Emissions from Power Plants in 2007*, Environment America, New York, http://cdn. publicinterestnetwork.org/assets/62514040f1134baf06e843fb233cd3ca/ EA_web.pdf

Alexander, R., Mazza, S. and Scholz, S. (2009) 'Measuring rates of return for lobbying expenditures: An empirical analysis under the Jobs Creation Act', University of Kansas School of Business, http://papers.ssrn.com/sol3/papers.cfm?abstract_id=1375082

Broder, J. (2010) 'New rules issued on coal air pollution', New York Times, 7 July, pA10

Government Accountability Office (2010) 'Surface coal mining: Financial assurances for and long-term oversite of mines with valley fills in four Appalachian States', GAO-10-206, Washington, DC, www.gao.gov/new.items/d10206.pdf

Lavelle, M. and Pell, M. (2009) 'The climate lobby from soup to nuts', Center for Public Integrity, www.publicintegrity.org/investigations/global_climate_change_lobby/articles/entry/1884/

Leiserowitz, A., Maibach, E., Roser-Renouf, C. and Smith, N. (2010) Global Warming's Six Americas: June 2010, Yale Project on Climate Change Communication, Yale University and George Mason University, New Haven, CT, http://environment.yale.edu/climate/files/SixAmericasJune2010.pdf

National Research Council (2009) Hidden Costs of Energy: Unpriced Consequences of Energy Production and Use, National Academy Press, Washington, DC

Packer, G. (2010) 'The empty chamber: Just how broken is the Senate?', New Yorker, 9 August

Pooley, E. (2010) The Climate War, Hyperion Books, New York

Repetto, R. (ed) (2006) Punctuated Equilibrium and the Dynamics of US Environmental Policy, Yale University Press, New Haven, CT

Repetto, R. (2007) 'The need for better internal oversight of corporate lobbying', Challenge, vol 50, pp76–96

Repetto, R. and Austin, D. (2000) Coming Clean: Financial Disclosure of Financially Significant Environmental Risks, World Resources Institute, Washington, DC

Repetto, R. and Dias, D. (2006) 'TRUEVA: The true picture', Environmental Finance, July–August

Science Daily (2008) 'Chronic illness linked to coal mining pollution, study shows', 27 March, accessed August 2010 at www.sciencedaily.com/releases/2008/03/080326201751.htm

Securities and Exchange Commission (2010) 'Interpretive guidance regarding disclosure related to climate change', Washington, DC, accessed 12 August 2010 at www.sec.gov/rules/interp/2010/33-9106.pdf

Chapter 7

Overcoming Obstacles to Adaptation

A FALSE SENSE OF SECURITY

The preceding chapters have explained the profound energy transition and policy reforms necessary to limit the rise in global temperatures to three or four degrees Fahrenheit, the increase agreed in the Copenhagen Accord as the maximum that is tolerable. Some have argued against taking action, claiming that the US can readily limit damages by adapting to climate change. This chapter comes to a much more pessimistic conclusion. Adaptation to climate change has lagged badly and faces serious obstacles. Depending only on adaptation to limit future damages would be reckless.

As climate changes, the risks of extreme weather damages rise disproportionately. Nonetheless, some influential studies have predicted that three or four degrees of warming would not be very damaging to the US and would even bring some benefits (Mendelsohn and Neumann, 1999; Nordhaus and Boyer, 2000). The main argument of these studies is that vulnerable organizations, firms and households will adapt. The US is rich in technology, economic resources, competent organizations and educated people, all of which create a high adaptive capacity. Moreover, the US economy has thrived in a wide variety of climatic conditions (Easterling et al, 2004).

The perception that US vulnerability is limited was used to rationalize the decision by the US Congress not to ratify the Kyoto Protocol on the grounds that the averted damages would be much smaller than the costs involved in reducing US greenhouse gas emissions to the extent being called for. Today, misunderstandings about our vulnerability still undermine support for national policies to reduce emissions. Consequently, assumptions regarding adaptation in the US have had broad policy repercussions.

Certainly, the potential for adaptation exists. Measures could be taken in all regions and sectors that could prevent or limit damages. Studies have indicated that the damages to agriculture, forestry and other economic activities can be greatly reduced if economic agents adapt efficiently (Mendelsohn and Neumann, 1999). Agricultural damages are estimated to be 50 per cent less if farmers adapt by changing cropping patterns, cultivation methods and irrigation practices (Mendelsohn and Neumann, 1999, p44). Economists have criticized the 'dumb farmer' assumption that farmers will just suffer damages and not do anything about it (Intergovernmental Panel on Climate Change, 2001). Other studies have estimated that damages from sea level rise could be reduced by over half, even including the costs of adaptation, if appropriate shoreline protective measures are taken (Neumann et al, 2000).

But finding that the US could adapt does not imply that it will adapt. The question is whether such steps will actually be taken and in sufficient time to limit damages. It doesn't follow that if Americans have adapted to a variety of climates in the past, they will be able to adjust to future climate changes. Some earlier mechanisms of adaptation will not be available in the future. For example, Americans adapted to semi-arid western environments by building irrigation systems and by tapping into underground aquifers. These options will not be available in the future if climate change dries out the region further, as scientists predict. In California, scarce water is already moving out of agriculture to serve more lucrative urban demands. The waters of the Colorado, the Rio Grande and other western rivers are already overcommitted. Reduced mountain snow packs and increased evaporation will reduce surface runoff into these rivers by as much as 40 per cent in parts of the Southwest by mid-century. Groundwater pumping has already depleted aquifers and lowered water tables in large parts of the West and the Great Plains. Water resources for more irrigation are not available.

Adaptation measures that anticipate and reduce vulnerabilities before damages are incurred are far better than reactive adaptation measures taken only after climate change damages have been suffered. If adaptation is mainly reactive, then damages will be much greater, because, as climate changes, responses will continually lag behind the rapidly rising risks. Unfortunately, experience shows that in the US, responses to vulnerabilities are mainly reactive. Airport security was tightened only after the terrorist attacks on the World Trade Center on September 11, 2001, although urgent warnings of danger had been made beforehand. Financial market regulations were tightened only after the disastrous financial collapse of 2007, although influential voices had been raised in warning beforehand. The barriers protecting New Orleans were strengthened only after catastrophic flooding occurred, although warnings of potential disaster had been circulating for years earlier. The safety of drilling for oil in the deep waters of the Gulf of Mexico was re-examined only after the Deepwater Horizon disaster, and only then were regulations tightened.

Even worse, whatever proactive adaptive measures are put in place may be partially or wholly offset by 'mal-adaptations' that increase vulnerability and the likelihood of future damages. One example is continuing shoreline and flood plain real estate development that increases the property values at risk. Development in the coastal zone has been the main factor in the rising trend in hurricane and coastal storm damage. The International Hurricane Research Center has identified Atlantic coastal regions in the US that are most vulnerable to future hurricane damage. In many of these coastal communities, rather than restricting development in the vulnerable beach areas, local governments are encouraging ongoing development by building sea walls and restoring beaches, a costly and flawed approach (Dean, 2006).

OBSTACLES TO ADAPTATION

The US has already experienced considerable climate change, in temperatures, precipitation, storm intensity and sea level rise. During the 20th century average temperatures in the US increased by 1.4 degrees Fahrenheit, with northern areas rising as much as 4 degrees Fahrenheit. The effects of those changes are already evident (National Assessment Synthesis Team, 2000). The fraction of precipitation that has fallen in intense weather events has increased, coastal wetlands have shrunk, intense hurricanes have become more frequent and coastal erosion has accelerated, especially in Alaska, where a decline in seasonal sea ice along the coasts has exposed shorelines to much more storm erosion. In the West, a larger fraction of winter precipitation is falling as rain; the snow season has shortened, reducing the annual snowpack; snowmelt is more rapid, changing river flows; and droughts are now more frequent (Barnett et al, 2008). Throughout the country, growing seasons and the ranges of some biotic organisms have shifted. Changes such as these over past decades are well documented (US Environmental Protection Agency, 2010).

Past increases in atmospheric concentrations of greenhouse gases have also already ensured that climate will change as much again in future decades, even if concentrations were miraculously stabilized overnight. Lags in the global climate system ensure that the full effects of any rise in concentrations are not felt for decades. Even worse, emissions are rising, atmospheric concentrations are increasing and climate change is accelerating. Even more damaging climate changes are likely in coming decades.

Lessons can be learned by examining adaptations that have been made to the substantial climate changes that occurred over the past half-century and to inevitable future changes. If adaptations have been thorough and rapid, then confidence is warranted that future adaptations may also be adequate. If not, then something can be learned about obstacles to adaptation that must be removed if future damages are to be limited.

Adaptation is sometimes inhibited by moral hazard situations. For example, government crop insurance and disaster relief programs reduce farmers' incentives to avoid weather damages. Between 1989 and 2009, indemnified losses insured by the Federal Crop Insurance Program, subsidized from the federal budget, increased at an average annual rate of 15 per cent, from $1.2 to $5.0 billion. During that period, the premium subsidy rose from $0.2 billion to $5.4 billion, an average annual increase of 39 per cent (Federal Crop Insurance Corporation, 2010). This subsidized insurance program provides strong incentives for farmers to increase their exposure to weather-related risks. The incentives are raised even further by farm subsidies that raise farmers' returns from agricultural operations whether or not crop damages occur. Federal disaster relief program payments have also risen rapidly over past decades, both absolutely and as a percentage of assessed damages. In recent years, even before Hurricane Katrina, these programs have covered roughly half of uninsured costs of weather-related disasters. Moreover, citizens, localities and State governments have come to expect federal disaster relief as an entitlement, accentuating the moral hazard problem (Moss, 1999).

Another obstacle arises because organizations are typically sluggish in responding to new conditions. The status quo is almost always the default decision-making option, and deviations from it are almost always relatively small and incremental. Drastic changes occur only very infrequently, usually in response to major threats or pressures (Baumgartner and Jones, 2002). Organizational actions and decisions are typically constrained by rules, routines, procedures, formulae and precedents. These are often codified, but even if not, they are enshrined as 'the way we do things' or 'best practice' and are usually difficult and costly to overturn and become badly out of date. Such routines are usually changed incrementally and reactively when existing ones are unsuccessful (Berkhout et al, 2008). But adaptations to new conditions will usually be drawn from a repertoire of already known actions, competencies and strategies. Therefore, most organizations change slowly and painfully, if at all.

Organizations' leaders and managers devote most of their attention and the organizations' resources to resolving immediate problems at the expense of longer term problems like climate change. The perception that climate change is an issue that can be put off and doesn't demand immediate attention makes it likely that it will not get attention. Neither organizations nor their leaders understand that climate change is steadily increasing the risks of weather disasters that might happen next month or next week.

The new field of behavioral economics has illuminated other characteristics of human decision-making under uncertainty that are likely to inhibit adaptation. Humans are myopic decision makers, sharply discounting events in the more distant future or past. People assign a relatively low priority to climate change, because its effects are not perceived to threaten the present. People also tend to underestimate cumulative probabilities when the probability of an event in a single period is low. For example, people build or buy houses in fire-prone, flood, mud slide and earthquake zones, even though the probability that an event will occur within their lifetimes is quite high.

People also exhibit strong 'anchoring' to the status quo, tending to make only small adjustments away from it. Many people, for example, refuse orders to evacuate when under immediate threat from natural disasters or move back into risky areas after disaster strikes. People also tend to resist and deny information that contradicts their values or ideological beliefs. Roughly ten per cent of Americans still are climate skeptics and continue to deny the scientific evidence regarding climate change and refuse to contemplate any changes in behavior.

Uncertainty is another major obstacle. Weather in most regions is notoriously variable, making inferences about long term changes difficult. Even trained scientists debated for decades whether climatic fluctuations exhibited any underlying trend. It is much more difficult for untrained observers to detect relatively small trends amid much larger short term fluctuations (Schneider et al, 2007). This is all the more true because people tend to place undue weight on recent events, extrapolating short term movements into the future. One or two cold years or even a big snowstorm is taken as evidence that there is no underlying warming.

Complicating matters further, climate typically varies much more in specific locations than when averaged over a large area. Over a large area, many short term fluctuations cancel out, but people form their judgments on local weather records, making it all the more difficult for people to discern trends. Moreover, because Americans are highly mobile, many people have not lived in the same locations for very long and can't see the trends.

Forecasting future climate change at small geographical scales is difficult even for climate scientists. Various global climate models might agree on global and broad, continental scale averages but differ in their predictions for smaller regions. Regional climate models derived from the global circulation models also differ in their predictions. Therefore scientists have been unable to provide firm guidance to local institutions.

As if these sources of uncertainty were not enough, people in and outside government who oppose action have deliberately sowed confusion and increased public doubt about climate trends. As the previous chapter on politics explains, major coal and oil companies such as Exxon Mobil provided financial support for years to 'climate skeptics', a small group of scientists who have persistently argued against the finding of the Intergovernmental Panel on Climate Change (IPCC) and other scientific bodies. More recently, there were attempts to discredit the entire 20-year effort of the IPCC, which is based on thousands of peer-reviewed studies and involves hundreds of scientists from scores of countries, because of trivial mistakes in one of its many reports.

THE RECORD ON ADAPTATION

How these obstacles affect adaptation is illuminated by examining how some of the most vulnerable organizations have responded to climate change risks, focusing on actions before 2008 by organizations that were especially likely to take adaptive measures because:

♦ their operations, investments or missions are vulnerable to climate change;

♦ they are making or planning long term fixed investments or long-running programs that will inevitably feel the effects of climate change; and

♦ they have the organizational capacity to forecast and plan.

In the public and quasi-private sectors, organizations that fit these criteria include land and forest management agencies that will be exposed to ecological and hydrological changes; flood control and disaster insurance agencies; water supply agencies in the West and Southwest that will feel the effects of drought and hydrological shifts; and transportation infrastructure agencies exposed to risks of sea level rise, storm surges and flooding. Have agencies that are making long term investments and commitments in all these vulnerable sectors changed their plans, designs, investment decisions, operational policies, budgets or staffing to reflect past or inevitable future climate change? If not, what obstacles have they faced and what should be done to accelerate future adaptation?

Hurricane damage

In the Gulf of Mexico, Hurricane Katrina created an exceptional storm surge that overwhelmed the levees protecting New Orleans, causing catastrophic flooding. In reviewing the Katrina disaster, Berkeley engineer Robert Bea discovered that the Corps of Engineers had applied a safety factor of 1.3 to the benchmark hundred-year flood height, estimated from historical data, in designing the height of levees (contrasted with a factor of 4–6 used in offshore oil platforms, which withstood the hurricane). This factor of 1.3 was carried over from the 1940s, when the Corps used it in building levees to protect agricultural land and pasture in the South from flooding. The levees the Corps had built were designed to allow up to two feet of water to overflow the barrier in a 100-year flood, as estimated with decades-old data (Schwartz, 2006). Bea also found that the Corps was rebuilding levees to the same standard – an earth mound without concrete sheathing – that had failed in the face of Katrina. This is a prime example of anchoring to past standards and practices.

Flood control

According to recent Congressional testimony from the Government Accountability Office, neither the National Flood Insurance Program (NFIP) nor the Federal Crop Insurance Corporation (FCIC) have developed information on their programs' longer term exposure to the potential risk of increased extreme weather events associated with climate change. Furthermore, according to NFIP and FCIC officials, both programs' estimates of weather-related risk rely heavily on historical weather patterns. As one NFIP official explained, 'the flood insurance program is designed to assess and insure against current, not future risks' (Stephenson, 2007). Only in 2009 did the Federal Emergency

Management Agency (FEMA) begin studies to re-estimate flood risks in order to adjust flood insurance premiums upwards to take into account the effects of climate change.

Real estate interests have lobbied against changes in the NFIP to update and expand flood zones within which flood insurance would be mandatory for homeowners with federally insured mortgages, because doing so would raise real estate costs. States where flooding has been infrequent have also opposed such updates, because most of the program's payouts have so far gone to just three States bordering the Gulf of Mexico. Established interests can and do oppose sensible adaptive measures.

The Corps of Engineers Flood Protection Program

In the Upper Mississippi Basin, a 1993 flood caused massive damage and scores of deaths, and subsequent floods have also been extremely damaging. The Army Corp of Engineers has partnered with other agencies to produce a comprehensive plan to coordinate and improve the more than 100 flood control systems in the basin. The underlying study used hydrologic frequency models based on historical records stretching back a century to design flood control structures and to estimate potential damages for insurance purposes (US Army Corps of Engineers, 2006). The study discussed the effect of climate change on future plans: 'Future climate change has the potential to change the frequency of flood events, manifesting itself as a shift in the discharge-frequency curve' (US Army Corps of Engineers, 2002). However, the study's draft final report did not factor climate change into its long term plan, which would have required revising hydrological models that had been codified in a painfully negotiated inter-agency agreement. According to the draft final report, 'for the purposes of this study, it is assumed that whatever climate changes occur within the 50-year planning timeframe will have little effect on the types of vegetation, cropping patterns or flood frequencies as currently determined'(US Army Corps of Engineers, 2006). This illustrates the power of codified rules.

Another obstacle stemmed from opposition within the Bush administration and Congress to steps to adapt water projects to climate change. For example, in 2007 the US Senate defeated a draft bill requiring the Army Corps of Engineers to consider the impact of climate change in designing water resource projects (Climate Change Capital, 2007).

Water supply

In the semi-arid Southwest more than 30 million people depend on water from the over-appropriated Colorado River Basin and from dwindling groundwater resources. More than a decade ago, the American Water Works Association Public Interest Advisory Forum recommended that planners assess the potential impacts of climate change on water supplies. The Association has published a primer on climate change for water utility managers (Miller and Yates, 2006). The Western Governors' Association also called for attention to climate change in water management (Western Governors' Association, 2006).

However, most of the necessary adaptation measures recommended by water experts still lie in the future. In 2007 water supply agencies in the Southwest and elsewhere joined forces to study how climate change is affecting the provision of drinking water in major metropolitan areas (Water Utility Climate Alliance, 2007). In 2006 the Western States Water Council (WSWC) recommended that regional, State and local water managers develop adaptation plans incorporating the changing probabilities of extreme climatic events. The WSWC acknowledged a continuing gap between researchers and water supply planners and managers:

Analytical uncertainties associated with assessing climate change impacts need to be addressed sooner rather than later, since results of those analyses are necessary early in the planning process. It thus makes sense to move expeditiously in developing the collaborative relationships with the climate research community that are important to procuring directed research outcomes. (Western States Water Council, 2007)

A follow-up assessment in 2010 found that the intervening years had been devoted mostly to requests for funding to expand data collection and for additional workshops. Much remains to be done to connect researchers with managers.

By 2007 water agencies in the Southwest were still not factoring climate change into plans and programs. For example, the Arizona Governor's Drought Task Force developed the Arizona Drought Preparedness Plan in 2004, which did not take climate change into account, because no consensus was reached on potential impacts. Also in 2004, the Colorado Water Conservation Board produced a Statewide Water Supply Investigation report projecting demands, availabilities and options through 2030. The report noted that these projections might be affected by climate change but neither used climate forecasts in projecting availabilities nor recommended using such forecasts in planning. In New Mexico, a State with declining groundwater levels and fully appropriated surface water resources, the 2003 State Water Plan published does not discuss the potential impacts of climate change. In Texas, where the western part of the State is largely dependent on groundwater, the Texas Water Development Board (TWDB) is constrained by law to use the historical 'drought of record' in water supply availability forecasts and has not incorporated climate change forecasts. Plans for the 16 Texas water regions are based on water availability models developed by the Texas Council on Environmental Quality and groundwater availability models developed by the TWDB, neither of which incorporate climate change impacts. Throughout the Southwest, where water availability is likely to be greatly affected by climate change within the planning horizon of current infrastructure investments, adaptation has been slow and tentative. These organizational responses continue to rely on historical rules and standards and to focus on current rather than future issues.

Land and natural resource management

Federal agencies are charged with sustainable management of over 600 million acres, almost 30 per cent of the US land area, as well as more than 150,000 square miles of protected waters. The principal management agencies are the Bureau of Land Management, the US Forest Service, the National Park Service and the National Oceanic and Atmospheric Agency, although other federal agencies, including the Department of Defense, also control considerable federal land.

In January 2001, the Secretary of the Interior directed these agencies to consider and analyze potential climate change effects in their management plans and activities. According to a 2007 Government Accountability Office (GAO) report, 'federal land and water resources are vulnerable to a wide range of effects from climate change, some of which are already occurring' (US Government Accountability Office, 2007). However, the 2007 report found that the agencies had not made climate change a priority and that the agencies' strategic plans did not specifically address climate change. Experts at a related GAO workshop identified challenges that confront resource managers. First, resource managers focus first on near term, required activities, leaving less time for addressing longer term issues such as climate change. Second, the maxim to 'let nature takes its course' has taken root, persuading many managers that taking no action is the appropriate response. Third, despite the January 2001 directive, the resource management agencies have not yet provided specific direction to managers in the field on implementing the order (Smith and Gow, 2007). Fourth, planning is based on past and current conditions and agencies lack the modeling capabilities to forecast the local and regional impacts of climate change. Uncertainty about ecological changes that climate change will produce and what, if anything, to do about them impedes adaptation.

Transportation infrastructure

In 2008, using the central Gulf Coast as a case study, the US Climate Change Science Program found substantial vulnerabilities to climate change in the nation's transportation infrastructure. For example, storm surges associated with hurricanes could easily reach seven meters in height, flooding more than half of the area's major highways, almost half of the rail mileage, 29 airports and almost all the ports. The assessment found that:

> Most agencies do not consider climate change projections per se in their long-range plans, infrastructure design or siting decisions. This appears to be changing, spurred in part by the devastating effects of Hurricanes Katrina and Rita. (US Climate Change Science Program, 2008)

Nonetheless, it was also found that 'none of the existing State and Metropolitan Planning Organization documents examined here, all of which date from 2000 to 2006, directly addresses or acknowledges issues of climate change and variability' (US Climate Change Science Program, 2008).

A study with similar conclusions by the Transportation Research Board of the National Academy of Sciences observed that:

Faced with a new problem such as this predicted break in trend, transportation professionals typically adopt incremental rather than radical solutions. This tendency to favor proven methods and practices is understandable, particularly for engineers, who are designing infrastructure expected to provide reliable service for decades, and in view of the uncertainties about the rate of climate change and the magnitude of its effects. Nevertheless, reinforced by conservative institutions, regulatory requirements and limited funding, this way of thinking can hamper timely responses to issues such as climate change that involve risk and uncertainty. (National Research Council, 2007)

Interviews conducted for the US Department of Transportation's Gulf Coast study by Cambridge Systematics, Inc. illustrate prevailing attitudes. When questioned about the possibility that climate change could bring about more storms of the intensity of Katrina or Rita, many local officials opted for a literal interpretation of federal planning guidance, which does not require consideration of climate change. Officials also pointed to federal policies that allow replacement of facilities only as they are currently designed, preventing consideration of design modifications that could provide for adaptation to potential climate change impacts. Transportation investments are guided by State and municipal plans that must conform to federal planning guidelines in order to make the investment projects eligible for federal funding. Those guidelines do not yet require consideration of climate change. Some officials even believed that federal regulations prevented them from considering any changes that would extend beyond the time horizon of their long-range plans. Still others identified funding shortfalls that, in combination with climate uncertainties and lack of relevant data, discourages planners from giving more attention to the issue (National Research Council, 2007). A follow-up report issued two years later revealed that uncertainties, lack of data and research funding still constrained adaptation (National Research Council, 2009).

RECENT DEVELOPMENTS IN THE NATIONAL GOVERNMENT

The Bush administration, during its eight years in office, opposed mitigation policies, resisted open communications on climate risks and did little to promote adaptation. For example, according to an inspector general's report at NASA:

> Our investigation found that during the fall of 2004 through early 2006, the NASA Headquarters Office of Public Affairs managed the topic of climate change in a manner that reduced, marginalized or mischaracterized climate change science made available to the general public. (Revkin, 2008)

The Obama administration has provided more leadership on adaptation. In December 2009 the Environmental Protection Agency issued a finding under the Clean Air Act that greenhouse gases endanger public health and welfare (US Environmental Protection Agency, 2009). In February 2010 the Council of Environmental Quality issued draft guidance for the preparation of environmental impact statements under the National Environmental Policy Act (NEPA), requiring agencies to assess the probable impact of foreseeable climate change on all significant proposed federal actions, especially on vulnerable long term projects. However, the main thrust of this guidance was to require assessment of potential GHG emissions and mitigation options (US Council on Environmental Quality, 2010a). In January 2009, the Secretary of the Interior again ordered agencies and bureaus within the department to consider potential climate change effects on management plans, research priorities and long term resource allocations. The EPA's Water Office has also developed a strategic plan for dealing with climate change.

However, federal activities still focus mostly on research, as the most recent report by the State Department to the United Nations Framework Convention on Climate Change makes clear (US Department of State,

2010). The Council on Environmental Quality now coordinates an inter-agency Climate Change Adaptation Task Force involving all major departments. Its progress report in March 2010 stressed the need for science inputs in decision-making, clearer communication of climate risks, coordination among agencies, capacity-building and prioritization of adaptation measures (US Council on Environmental Quality, 2010b). Even after a half-century of climate change, national leadership on adaptation is still at an early stage.

STATE AND LOCAL LEVEL ADAPTATION RESPONSES

A recent review by the Pew Center on Global Climate Change reported that only ten coastal States prepared or were preparing climate adaptation plans (Cruze, 2009). In 27 other States, climate change action plans were under preparation but less than half include any substantial mention of adaptation planning. Even plans that did so discussed future, not ongoing actions. By contrast, many county and municipal governments are planning or taking action to reduce their vulnerabilities to climate change damages. ICLEI, Local Governments for Sustainability, has been stimulating these activities through its Climate Resilient Communities Initiative. However, as at the State level, most community climate change action plans focus on mitigation options, not adaptation, or seek to identify adaptation measures that could be taken in the future.

No State needs to initiate adaptation measures more urgently than Alaska, where temperature has already risen over the last century by 3.5 degrees Fahrenheit (and by 6 degrees during winter). Alaska is already experiencing significant climate change impacts, including infrastructure damage from melting permafrost, dislocation of more than 100 coastal communities from shoreline erosion as sea ice barriers disappear, increasingly severe and widespread forest fires, forest pest outbreaks, and changes in fish and marine mammal populations. Infrastructure damages alone are expected to add between $3.5 and $6 billion to normal maintenance and replacement costs over the next two decades.

Scientists at the University of Alaska have created climate forecasts for the State showing dramatic future warming – as much as 20 degrees Fahrenheit in winter temperatures. Nonetheless, the 2008 Alaska Climate Impact Commission's final report to the State Legislature found only outdated and deficient data with which to plan for adaptation. Shoreline and maritime maps, precipitation frequency distributions, and census data on more than 16,000 items of infrastructure and hundreds of communities at risk need to be updated for planning purposes. The Alaska Department of Transportation, for example, reported to the Commission that it 'has not systematically studied the need for, or implemented, specific changes to policy or regulations relative to climate change, nor does it have pertinent data upon which to base such changes'. With respect to adaptation responses, 'generally, the Commission feels that we are early in the period of climate change understanding when it comes to determining precise budget impacts and service delivery changes by State government'. The Commission's recommendations were only to remedy this planning deficit (Alaska State Legislature, 2008).

PRIVATE SECTOR RESPONSES

Many private businesses are exposed to significant climate risks to their own facilities, infrastructure and staff as well as to those of their suppliers and customers. It has been presumed that the private sector will adapt efficiently and briskly to those risks: 'Efficient private adaptation is likely to occur, even if there is no official (government) response to global warming' (Mendelsohn and Neumann, 1999).

Some private sector organizations are adapting to changing climate conditions, albeit mostly reactively. In the aftermath of Katrina, oil companies with offshore and onshore facilities in the Gulf responded to studies indicating that climate change will produce more intense hurricanes in the region by revising their design standards for offshore oil rigs and pipelines. Companies have already seen presumably one in a hundred year storms happening every few years (Mouawad, 2005).

Insurance companies are concerned about increasing storm losses, especially from hurricane damages (Emanuel, 2005 and 2006). Some have reduced coverage in vulnerable areas. Many are re-examining their actuarial estimates and have markedly increased premium rates. Yet, according to one observer, 'Although insurers first expressed concern about climate change more than three decades ago, fewer than one in a hundred appear to have seriously examined the business implications' (Mills, 2005). Efforts by the insurance companies to project future hurricane losses through quantitative risk modeling have been obstructed in some States, including Texas and Florida, by insurance regulatory commissions that have recommended against the use of such models in rate-making.

A review by the consulting firm KPMG identified industries that perceive themselves to be at greatest physical risk from climate change (healthcare, agriculture and forestry, transportation, insurance, and tourism) as well as others at considerable risk (real estate, finance, construction and materials, retail, and manufacturing). The review found that in the healthcare, tourism and transportation industries there is generally a low degree of preparedness or adaptation to the physical risks of climate change, despite a high perceived level of exposure. In the retail, financial and real estate sectors, they found only a moderate degree of preparedness (KPMG International, 2008).

A complementary study by the Pew Center on Global Climate Change also found that:

> The physical risks of climate change are often overlooked by business. The reasons for this are several: the uncertainty of future projections

and the long-term nature of the change make it easy for businesses to set aside current climate risk, and concerns about greenhouse gas emissions and mitigation are more pressing to corporate leaders and shareholders. Moreover, many decision-makers have yet to recognize that the past is not the best predictor of the future – whether for climate averages or climate variability. (Sussman and Freed, 2008)

UNDERESTIMATING EXTREME WEATHER RISKS

One of the toughest obstacles to adaptation is that the severest climate damages will be caused by extreme weather events: unusual heatwaves, droughts, floods, hurricanes and storm surges. Organisms and ecosystems can tolerate a range of weather conditions, and man–made structures and systems are designed to do so as well. Within this range of tolerance, weather variability causes little damage and if change is sufficiently gradual, many systems can adapt or be adapted. When weather varies outside this range of tolerance, however, damages increase disproportionately. As floodwaters rise, damages are minimal as long as the levees hold, but when levees are overtopped, damages can be catastrophic. If roofs are constructed to withstand 80mph winds, a storm bringing 70mph winds might only damage a few shingles, but if winds rose to 100mph, roofs might come off and entire structures be destroyed. Plants can withstand a dry spell with little loss of yield, but a prolonged drought will destroy the entire crop.

Even when the television news dramatizes unprecedented heatwaves, forest fires and flooding, few people associate these disasters with climate change. Some think they're acts of God; others, just bad luck that comes along occasionally. Even scientists have difficulty. Because extreme events are infrequent, estimating their probability is difficult, because there are so few observations in the historical record. In estimating the probability of events occurring infrequently, at the 'tails' of the probability distribution, the underlying probability distribution that is adopted to represent the data is crucial, because it will estimate the likelihood of extreme events when extrapolated beyond the range of existing data. Moreover, since there will be very few, if any, observations or data points representing these extremes, sampling error in fitting the distribution to the historical record will be large.

More fundamentally, distributions estimated from historical data are increasingly unrepresentative of future conditions as climate change continues. Even if weather conditions don't become more volatile, which might happen, a shift in average conditions will bring about a rapidly changing probability of weather events far removed from average conditions (Nogav et al, 2007). For example, as more rain falls in heavy storms, the probability rises of occasional deluges that bring about extreme flooding. As average temperatures rise, the likelihood of an extreme heatwave rises too.

However, it's difficult to judge when the probability of extreme events has changed. The flooding in Iowa in June 2008 had not been experienced since 1851, if then. Does that flood signal that severe flooding has become more likely or was it merely a very unlikely event? It might take decades and several occurrences to conclude with statistical certainty that what had been regarded as a 'once-in-a-hundred-year flood' has become a 'once-in-fifty-year flood'. Most people can't deal with these complexities.

These uncertainties are inhibiting adaptation even more, because the risks of weather-related disasters are being significantly underestimated by government agencies charged with estimating the likelihood of extreme weather events. These are the scientific bodies that should be providing leadership, but in estimating risks from hurricanes, floods, droughts and other extreme weather, they have not yet adequately factored in the effects of past and future climate change. The frequencies with which specific weather events occur are estimated from measurements in the historical record going back decades. These frequencies, calculated from past records, are then used to 'fit' to the data a probability distribution with a similar mean, variance and skewness. The probability distribution can then be used to estimate the likelihood of extreme weather, such as a scorching hot day, even though there are few, if any, such events in the historical record.

When climate is changing, it's highly misleading to assume that the future will be like the past and then to project probabilities estimated from historical data into the future (Milly et al, 2008). Not only are agencies charged with assessing weather risks making this assumption, they are also ignoring measured trends in historical weather patterns. They do so because of uncertainty as to whether an apparent trend is real or is just a poorly understood cyclical phenomenon that will be reversed, or just a string of random events.

Agencies cling to conservative methodologies and outdated estimates almost certain to be erroneous rather than make use of scientific projections of future conditions that are still uncertain, especially at regional or local geographic scale. The question bedeviling weather risk assessment is 'If the future will not be like the past, what will it be like?'. Without firmer guidance from climate scientists or policy decision makers, government risk assessors are reluctant to make a stab at answering this question.

Weather risk assessments become increasingly outdated as time passes or when projected further into the future. They provide unreliable guidance for the design, placement and construction of infrastructure that will be in place for many decades and vulnerable to extreme weather. By underestimating future risks, they also provide unreliable guidance for investment and program decisions to adapt existing infrastructure and communities to more extreme weather. As a result, according to a 2009 report by a National Research Council panel:

> Government agencies, private organizations and individuals whose futures will be affected by climate change are unprepared, both conceptually and practically, for meeting the challenges and opportunities

it presents. Many of their usual practices and decision rules – for building bridges, implementing zoning rules, using private motor vehicles and so on – assume a stationary climate – a continuation of past climatic conditions, including similar patterns of variation and the same probabilities of extreme events. That assumption, fundamental to the ways people and organizations make their choices, is no longer valid. (National Research Council, 2009)

Government and private organizations relying on these outdated risk assessments have consequently invested too little to adapt existing and newly constructed infrastructure to the effects of changing climate. New investments designed to outmoded risk standards are likely to suffer excess damages and poor returns. This is a problem of broad and significant scope. Among the public and private sector organizations that are exposed to increasing but underestimated risks are:

- local, State and federal disaster management agencies;
- local, State and federal agencies that finance and build public infrastructure in vulnerable areas as well as those that own and operate vulnerable infrastructure;
- private investors and owners of vulnerable buildings and other physical property;
- property and casualty insurers;
- creditors holding vulnerable infrastructure directly or indirectly as collateral; and
- vulnerable businesses and households.

Clearly, this listing encompasses a large proportion of the American economy, and an assessment of the vulnerable regions would extend over much of the country, including coastal regions subject to hurricanes, storm surges and erosion; river basins subject to flooding; and agricultural areas subject to wind, storm and drought damage.

CASE STUDY: HURRICANE RISK IN THE NEW YORK CITY REGION

Consider the increasing risks to the New York City metropolitan region from hurricane damage. The New York metropolitan region extends across three States and encompasses an extraordinarily dense concentration of infrastructure, physical assets and business activity. In 2006 the value of insured coastal property in the New York, Connecticut and New Jersey metropolitan region was almost $3 trillion (Valverde, 2006). The region is vulnerable to hurricanes. Storm surges could reach 18–24 feet in a strong hurricane. Low-lying regions, including Kennedy Airport and Lower Manhattan, would flood. Roads, subway and tunnel entrances would be submerged, along with ground level and underground infrastructure. High winds would do severe damage, partly by blowing dangerous debris through city streets.

Regional risk assessment starts with the probability assessment carried out by the National Hurricane Center (NHC) within the National Oceanic and Atmospheric Administration. The methodology used for New York City and other coastal regions counts the occurrence of hurricanes of specific intensities (defined in terms of maximum sustained wind speeds) striking within a 75-mile radius during the historical record of approximately 100 years. NHC scientists fitted a Weibull probability distribution to these observed frequencies and the probabilities of hurricanes of various intensities were then read off the fitted probability distribution. There were no actual observations of the most severe hurricanes in the historical record for the New York region, so those probabilities were extrapolations based on the fitted distribution. The results, expressed as the expected return periods, are shown in the column for 2000 in Table 7.1 for various categories of hurricanes (National Hurricane Center, 2010). The return periods are reciprocals of annual probabilities: for example, a return period of 39 years implies a 0.026 annual probability.

The NHC probability estimates were constructed in 1999 and were probably not valid even then, because there has been an upward trend in intense hurricanes in the North Atlantic over at least the past 35 years. The number of Category 4 and 5 hurricanes in the North Atlantic increased from 16 during the period 1975–1989 to 25 from 1990–2004, so the earlier years in the historical record used to compute frequencies might not have been representative of the final years (Webster et al, 2005).

Even worse, the methodology takes no account of the effects of climate change. The National Hurricane Center calculations of return periods were still based on 1999 data as of May 2010. According to the guidelines still used to calculate probabilities:

> There is much speculation about climatic changes. Available evidence indicates that major changes occur in timescales involving thousands of years. In hydrologic analysis it is conventional to assume that flood flows are not affected by climatic cycles or trends. Climatic time invariance was assumed in developing this guide. (US Geological Survey, 1982)

The increasing frequency of stronger hurricanes in the North Atlantic is probably linked to climate change through the gradual rise in sea surface temperatures (Union of Concerned Scientists, 2006). Warming ocean waters provide the energy that creates and sustains more intense hurricanes (Emanuel, 1987). According to a recent study, a three degree centigrade increase in sea surface temperature would raise maximum hurricane wind speeds by 15 to 20 per cent (Sriver and Huber, 2006).

Sea surface temperatures are rising at a rate of approximately 0.14 degrees centigrade per decade (Casey and Cornillon, 2001). The warming rate is apparently increasing, however, and the North Atlantic warming exceeds the global average. From 1981 to 2009 warming in the North Atlantic has averaged 0.264 degrees centigrade per decade, roughly twice the global average (Tisdale, 2009). Because of rising sea surface temperatures in the North Atlantic, the driving force behind the increasing frequency of intense hurricanes, backward-looking probability estimates generated using the National Hurricane Center's approach do not adequately reflect current and future risks. This problem is compounded by the rising sea level, also partly the result of increasing ocean temperatures. In the North Atlantic between New York and North Carolina, sea level has risen more rapidly than the global average, at rates between 0.24 and 0.44 centimeters per decade (US Climate Change Science Program, 2009). Higher sea levels and tides make flooding driven by hurricane force winds more probable.

The trend in sea surface temperature and its relationship to maximum wind speed provides a way to forecast changes in the intensity of future hurricanes. Though forecasts based on this approach are uncertain, high and low estimates can define a range of future probabilities and the results are more useful than estimates that incorporate no relevant information about the effects of climate change.

Table 7.1 shows estimated return periods for hurricanes striking the New York region in 2010, 2020 and 2030, estimated by linking the National Hurricane Center's probability distribution to a time trend based on the rate of sea surface temperature change and its effect on maximum wind speeds. The ranges shown for the decades 2010–2030 reflect the high and low estimates of the rate of sea surface temperature increase. The table shows that climate change will increase the probability of hurricanes striking New York, especially the more severe hurricanes. By 2030, the probabilities of Category 4 and Category 5 hurricanes striking the New York metropolitan region are likely to have increased by as much as 25 and 30 per cent respectively. These changing probabilities have dramatic economic implications.

Table 7.1 Estimated hurricane return periods, 2000–2030

Hurricane category	2000	2010	2020	2030
1	17	13.9–14.1	13.5–13.9	13.1–13.7
2	39	41.8–42.7	39.9–41.6	38.1–40.6
3	68	63.9–65.8	60.1–63.6	56.6–61.5
4	150	136.7–142.2	125.9–135.8	116.2–129.9
5	370	327.2–346.1	290.7–324.2	259.2–304.1

The economic implications of increasing weather risk

A professional risk management consultancy estimated in 2006 that a Category 3 hurricane striking the New York metropolitan region would cause approximately $200 billion in property damage, business losses and other impacts (Clark, 2006). According to the National Hurricane Center's estimates for 2000, there is only a 1.5 per cent annual probability of that happening. A more complete assessment makes use of the loss exceedance curve, commonly used in the insurance industry to represent the annual probability of a loss equal to or greater than specified amounts. A loss exceedance probability of $200 billion represents the chance that a hurricane loss might occur of that amount or more, into the trillions of dollars. Constructing a loss exceedance curve for the New York region requires not only the probabilities of Category 1–5 hurricanes but also the damages that they would inflict.

A recent study by Yale economics professor William Nordhaus, based on hurricanes recorded throughout the US, investigated the relationship between maximum wind speeds and resulting damages (Nordhaus, 2006). Shockingly, this study found that damages increase as the eighth power of the wind speed: if a hurricane with wind speeds of 50 miles per hour would cause $10 billion in damages, then one with maximum winds of 100 miles per hour would cause not twice the damages, $20 billion, but more than $2500 billion. The reasons for this dramatic escalation are threefold. First, higher winds will obviously do more damage to everything in their path; second, more intense hurricanes are likely to have impacts over wider areas; and third, their winds are likely to persist at damaging speeds, although not at the maximum, for longer periods of time.

The loss exceedance curve implied by this relationship is plotted in Figure 7.1 for 2000 and for subsequent decades, using the higher estimate of sea surface temperature increase. On the horizontal axis, damages are marked in hundreds of billions of dollars. On the vertical axis are the probabilities of hurricane losses of those amounts or more. One striking feature is that the curve is 'fat-tailed': probabilities decline slowly as heavy losses mount. The probability of losses exceeding a trillion dollars is substantially more than half the probability of losses exceeding $500 billion. This illustrates how vulnerable to catastrophic hurricane damage New York is now.

Figure 7.1 also shows that the probabilities of large losses are increasing over time. By 2030, the probability of hurricane damages exceeding amounts in the range of $100 to $500 billion could be 30–50 per cent greater than current estimates assume. Warming sea surface temperatures and rising sea levels are increasing the economic risks to coastal cities from intense hurricanes. In the absence of effective mitigation and adaptation measures, risks of catastrophic losses will continue to rise over coming decades.

Because risks are increasing, even without an extra aversion toward catastrophic risks, the region should be willing to pay an annual insurance premium up to the expected value of damages, if comprehensive insurance were available. The expected value of losses is the sum of all possible hurricane losses, weighted by their probabilities. That insurance premium, calculated using the outdated 2000 return periods estimated by the National Hurricane Center, is about $33 billion dollars per year.

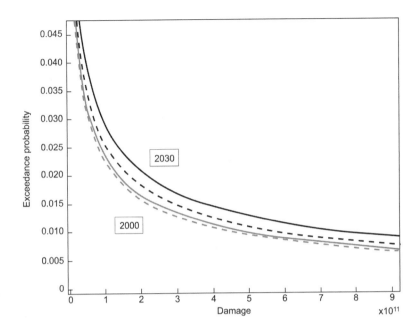

Figure 7.1 *Hurricane loss exceedance curves, 2000–2030*

However, it increases to $35–37 billion in 2010, $40–46 billion in 2020 and $47–62 billion in 2030, nearly doubling over three decades without inflation due only to the increasing likelihood of intense hurricanes. Unfortunately, the reality is even more disturbing. William Nordhaus's investigation and others have found that damages increase over time for the same maximum wind speeds, because of increasing population and infrastructure in the coastal zone. Were vulnerabilities to increase over coming decades as fast as in the past, a rate that Nordhaus estimated at 2.9 per cent per year, by 2010 the expected value of annual damages would have increased to $45–47 billion, by 2020 they would be $65–75 billion and by 2030 they would range from $100 to $130 billion.

Risks to investors

Investors in buildings, roads or other structures vulnerable to hurricane damage are likely to experience rates of return from their investments dramatically below those that they anticipate. Infrastructure projects are designed to withstand extreme weather, but there is a tradeoff between an extra margin of safety and the additional cost required to achieve it. Civil engineers and planners base decisions on such tradeoffs, but unfortunately they are still relying on historical frequency estimates and assuming that the future will be like the past, despite climate change. Thought leaders in the engineering profession have only recently begun weighing alternative approaches to climate change issues (US Geological Survey, 2008). Most practicing engineers still adhere to 'best practice', using assumptions validated by past experience, not innovative approaches aimed at new challenges.

For example, consider a new infrastructure project with a 40-year lifetime, designed to meet the hurricane risks calculated in 1999. The investor might require an expected income stream that would provide a discounted present value return of 12 per cent on the investment, taking into account possible income losses from hurricane damage. Unfortunately, as the years pass, the initial risk estimates are no longer valid. The expected returns on the project are dramatically affected, as Table 7.2 illustrates. These results are based on three alternative discount rates: three, five and eight per cent. Higher discount rates give less weight to the higher risks of hurricane damage in future years.

Hurricane damages are estimated as a function of maximum wind speed, but the more conservative Carvill index is used, which relates damages to the third power of wind speed rather than to the eighth power, because it underlies some recent financial derivative instruments available to hedge hurricane risk. Moreover, the calculation assumes that hurricane damages sustained in any year are limited to that year, as if they could be immediately repaired. No damages are assumed for wind speeds below 30 miles per hour. A total loss is assumed for speeds above 130 miles per hour, and at intermediate wind speeds damages rise according to the Carvill index. Despite these conservative assumptions, Table 7.2 shows that the project is unlikely to earn nearly the planned 12 per cent return. At a 3 per cent discount rate, expected investment returns would be reduced by almost 90 per cent, with a significant probability that the project would not repay the capital investment. The message is clear: designing vulnerable infrastructure projects without adequately estimating future weather risks will lead to significant investment losses.

Table 7.2 *Expected return on a project investment with increasing climate risk*

Discount factor	Expected rate of return
0.03	1.3%
0.05	3.9%
0.08	6.7%

Investments in adaptation and prevention

Not surprisingly, the economic value of investments in adaptation and prevention are also dramatically underestimated by using historical risk measures. With the previous investment analysis as a starting point, imagine that at an additional investment cost, the structure can be strengthened to withstand an additional 10mph of maximum wind speed without any additional damage. The payoff to this adaptation investment would be a lower risk of hurricane damage and a higher expected income return. Suppose further that such an investment in adaptation would just break even if the historical hurricane frequencies were valid over the project's anticipated lifetime. Under these assumptions, adaptation would be considered uneconomic, since it would yield no positive return on investment.

If the increasing probabilities of more extreme storms striking the region were taken into account, then investing in adaptation would appear much more attractive. Table 7.3 shows the expected returns on such an investment if the future effects of climate change are considered. The differences are dramatic: at a 3 per cent time discount rate, the zero return on an adaptation investment rises to a 68 per cent return.

Table 7.3 *Returns to an adaptation investment with increasing risk*

Time discount rate	Return assuming historical risk	Returns assuming changing risks
0.03	0%	68%
0.05	0%	56%
0.08	0%	43%

Since, with few exceptions, private investors and public agencies at local, State and federal levels are still relying on static, historically based probability estimates of extreme weather events and have not yet incorporated the effects of climate change into these risk estimates when evaluating the economics of adaptation investments, these agencies are grossly underestimating the economic case for investments in adaptation. This is one of the reasons why adaptation has lagged and is proceeding so slowly.

New York City municipal authorities have begun to plan adaptation strategies. There is much to do. The city's building codes are 40 years old. With respect to wind damages, they require only protection up to 110mph winds. With respect to flooding, they rely on 1983 flood maps corresponding to a Category 1 hurricane and are based on historical data. Even the newer replacement maps adopted in late 2007, with enlarged flood zones, are still based only on historical data (Rogers, 2008). In August 2008 mayor Bloomberg created the NYC Climate Change Adaptation Task Force, comprised of dozens of agencies, private companies and scientists. Its tasks include assessing vulnerabilities, developing adaptation strategies and design guidelines. Its report is due in 2010.

CONCLUSIONS

The US is increasingly vulnerable to severe hurricanes, windstorms, floods and droughts that would cause enormous damage to communities and their physical assets. With few exceptions, even after decades of documented climate change, vulnerable organizations are at early stages of research to develop adaptation strategies. There have been very few changes in forecasts, plans, design criteria, investment decisions, budgets or staffing patterns in response to climate risks.

Private and public sector organizations face significant obstacles to adaptation: uncertainty regarding future climate change at regional and local scales; uncertainty regarding the future frequency of extreme weather events; and uncertainty regarding the ecological, economic and other impacts of climate change. Organizations lack relevant data for planning and forecasting, and existing data are typically outdated and unrepresentative of future conditions. Organizations also face institutional and human barriers to adaptation: the need to overcome or revise codes, rules and regulations that impede change; the lack of clear directions and mandates to take action; political or ideological resistance to the need to respond to climate change; the preoccupation with near term challenges and priorities; the lingering perception that climate change is a concern only for sometime in the future; and the inertia created by a business as usual assumption that future conditions will be more or less like those of the past.

To say that the US can adapt to climate change does not imply that the US will adapt. Without strong national leadership and concerted efforts to remove these barriers and obstacles, adaptation to climate change will continue to lag. It will be largely reactive rather than anticipatory and preventive. Federal, State and local governments should begin at once to inventory and prioritize vulnerabilities. They should identify laws, regulations, incentives and programs that inhibit adaptation and then change them. They should require that design standards, building codes and planning guidelines be based on the best available projections of future climate conditions, not those of the past. Though the effects of climate change on local and regional weather patterns are still uncertain, uncertainty doesn't justify paralysis. It should be incorporated into estimates of future risks by establishing plausible ranges for key variables and parameters. Adhering to estimates almost certain to be wrong, waiting for uncertainties to be resolved, provides misleading information for current decisions. The dangerous effects of climate change must be prevented.

Most of all, neither public policy nor private action should be based on the hope that future adaptation can take the place of strong immediate measures to reduce emissions and halt the rise of greenhouse gas concentrations in the atmosphere. That unrealistic hope would expose the country to a worsening series of weather disasters.

REFERENCES

Alaska State Legislature (2008) *Alaska Climate Impact Assessment: Final Commission Report*, Juneau, AL

Barnett, T., Pierce, D., Hidalgo, H., Bonfils, C., Santer, B., Das, T., Bala, G., Wood, A., Nozawa, T., Mirin, A., Cayan, D. and Dettinger, M. (2008) 'Human-induced changes in the hydrology of the Western US', *Science*, vol 319, pp1080–1083

Baumgartner, F. and Jones, B. (2002) *Policy Dynamics*, University of Chicago Press, Chicago, IL

Berkhout, F., Hertin, J. and Gann, D. (2008) *Learning to Adapt: Organizational Adaptation to Climate Change Impacts*, Tyndall Centre for Climate Change Research, Oxford, UK

Casey, K. and Cornillon, P. (2001) 'Global and regional sea surface temperatures', *Journal of Climate*, vol 14, pp3801–3818

Clark, K. (2006) 'Major hurricane strikes the Northeast: How large will the losses be?', presentation to the Northeast Hurricane Conference AIR Worldwide, New York, 19 July

Climate Change Capital (2007) *Climate Change News Round-up*, London, UK

Cruze, T. (2009) *Adaptation Planning – What US States and Localities are Doing*, Pew Center on Global Climate Change, Washington, DC

Dean, C. (2006) 'Next victim of warming: The beaches', *New York Times*, 20 June

Easterling, W., Hurd, B. and Smith, J. (2004) *Coping with Climate Change: The Role of Adaptation in the US*, Pew Center on Global Climate Change, Washington, DC

Emanuel, K. (1987) 'The dependence of hurricane intensity on climate', *Nature*, vol 326, pp483–485

Emanuel, K. (2005) 'Increasing destructiveness of tropical cyclones over the past 30 years', *Nature*, vol 436, pp686–688

Emanuel, K. (2006) 'Climate and tropical cyclone activity: A new model downscaling approach', *Journal of Climate*, vol 19, pp4797–4802

Federal Crop Insurance Corporation (2010) 'Summary of business reports, 1989–2008', Washington, DC

Intergovernmental Panel on Climate Change (2001) *Working Group II: Impacts, Adaptation and Vulnerability*, Geneva, Switzerland

KPMG International (2008) 'Climate changes your business', www.kpmg. com/Global/en/IssuesAndInsights/ArticlesPublications/Pages/Climate-changes-your-business.aspx

Mendelsohn, R. and Neumann, J. (1999) *The Impact of Climate Change on the US Economy*, Cambridge University Press, New York

Miller, K. and Yates, D. (2006) *Climate Change and Water Resources: A Primer for Municipal Water Providers*, American Water Works Association, Denver, CO

Mills, E. (2005) 'Insurance in a Climate of Change', *Science*, vol 308, pp1040–1044

Milly, P., Betancourt, J., Falkenmark, M., Hirsch, R., Kundzewicz, Z., Lettenmaier, D. and Stoufferet, R. (2008) 'Stationarity is dead: Whither water management?', *Science*, vol 319, pp573–574

Moss, D. (1999) 'Courting disaster? The transformation of federal disaster policy since 1803', in K. Froot (ed) *The Financing of Catastrophe Risk*, University of Chicago Press, Chicago, IL

Mouawad, J. (2005) 'At time of epic storms, the oil industry thinks again', *New York Times*, 15 September

National Assessment Synthesis Team (2000) *Climate Change Impacts in the US*, Cambridge University Press, New York

National Hurricane Center (2010) 'Hurricane return periods: Northeast Region', accessed August 2010 at www.nhc.noaa.gov/HAW2/english/basics/return/shtml

National Research Council (2007) *Potential Impacts of Climate Change on US Transportation*, Transportation Research Board, Special Research Report 290, National Academy Press, Washington, DC

National Research Council (2008) *Informing Decisions in a Changing Climate*, Panel on Strategies and Methods for Climate-Related Decision Support, National Academy Press, Washington DC, pS1

National Research Council (2009) *A Transportation Research Program for Mitigating and Adapting to Climate Change and Conserving Energy*, Transportation Research Board, Special Research Report 299, National Academy Press, Washington, DC

Neumann, J., Yohe, G. and Nicholls, R. (2000) *Sea Level Rise and Global Climate Change: A Review of Impacts on US Coasts*, Pew Center on Global Climate Change, Washington, DC

Nogav, M., Parey, S. and Dacunha-Castelle, D. (2007) 'Non-stationary extreme models and a climate application', *Nonlinear Processes in Geophysics*, vol 14, pp305–316

Nordhaus, W. (2006) 'The economics of hurricanes in the US', National Bureau of Economic Research Working Paper W12813, http://papers.ssrn.com/sol3/papers.cfm?abstract_id=955246

Nordhaus, W. and Boyer, J. (2000) *Warming the World*, MIT Press, Cambridge, MA

Revkin, A. (2008) 'NASA office is criticized for climate reports', *New York Times*, 3 June

Rogers, T. (2008) 'How safe is my home?', *New York Times*, 1 May

Schneider, S., Easterling, W. and Means, L. (2007) 'Adaptation: Sensitivity to natural variability, agent assumptions and dynamic climate change', *Climate Change*, vol 45, pp203–221

Schwartz, J. (2006) 'The dilemma of the levies: New data, conflicting requirements and outdated standards at the Corps', *New York Times*, 1 April

Smith, M. and Gow, F. (2007) 'Unnatural preservation', *High Country News*, February

Sriver, R. and Huber, M. (2006) 'Low frequency variability in globally integrated tropical cyclone power dissipation', *Geophysical Research Letters*, vol 33, pL11,705

Stephenson, J. (2007) 'Climate change: Financial risks to federal and private insurers in coming decades are potentially significant', Statement of the Director of Natural Resources and Environment, Government Accountability Office, before the House Select Committee on Energy Independence and Global Warming, Washington, DC, 3 May

Sussman, F. and Freed, J. (2008) *Adapting to Climate Change: A Business Approach*, Pew Center on Global Climate Change, Washington, DC

Tisdale, R. (2009) 'The impact of the North Atlantic and volcanic aerosols on short-term global sea surface temperature trends', http://bobtisdale.blogspot.com/2009/02/impact-of-north-atlantic-and-volcanic.html

US Army Corps of Engineers (2002) 'Uncertainty of flood frequency estimates: Examining effects of land use changes, climate variability and climate change: Synthesis report', Washington, DC, 31 October, p10

US Army Corps of Engineers (2006) 'Upper Mississippi comprehensive plan: Draft report', Washington, DC

US Climate Change Science Program (2008) 'Impacts of climate change and variability on transportation systems and infrastructure: Gulf Coast Study, Phase 1; Executive Summary', Washington, DC

US Climate Change Science Program (2009) 'Coastal sensitivity to sea level rise: A focus on the mid-Atlantic region: Synthesis and Assessment Product 4.1', Washington, DC

US Council on Environmental Quality (2010a) 'Draft guidance on the consideration of greenhouse gases', Executive Office of the President, Washington, DC, 18 February

US Council on Environmental Quality (2010b) 'Climate change adaptation task force progress report', Executive Office of the President, Washington, DC, 16 March

US Department of State (2010) 'US Climate Action Report (2010): Fifth national communication of the US of America under the United Nations Framework Convention on Climate Change', Washington, DC

US Environmental Protection Agency (2009) 'Endangerment and cause or contribute findings for greenhouse gases under Section 202a of the Clean Air Act', Washington, DC, 9 December

US Environmental Protection Agency (2010) 'Climate change indicators in the US', Washington, DC, April

US Geological Survey (1982) 'Bulletin 17b. Guidelines for determining flood frequencies', Washington, DC, http://water.usgs.gov/osw/bulletin17b/bulletin_17B.html

US Geological Survey (2008) *Climate Change and Water Resources: A Federal Perspective*, Washington, DC

US Government Accountability Office (2007) 'Climate change: Agencies should develop guidance for addressing the effects on federal land and water resources: Summary', Washington, DC, 7 August

Union of Concerned Scientists (2006) *Hurricanes in a Warmer World*, Cambridge, MA

Valverde, J. (2006) 'Hurricane risk in New York City and Long Island', Insurance Information Institute, New York, accessed at www.iii.org/media/presentations/hurricaneriskny

Water Utility Climate Alliance (2007) www.wucaonline.org/html/about-us. html

Webster, P., Holland, G., Curry, J. and Chang, H. R. (2005) 'Changes in hurricane number, duration and intensity in a warming environment', *Science*, vol 309, pp1844–1846

Western Governors Association (2006) *Water Needs and Strategies for a Sustainable Future*, Denver, CO

Western States Water Council (2007) 'Proceedings of the Climate Change Research Needs Workshop', Denver, CO

Chapter 8
Summary and Conclusions

The world is standing on the brink of the worst environmental disaster in hundreds and possibly thousands of years. Over the past century, reliable measurements have shown that Earth is warming, sea surface temperatures and the sea level itself have risen, Arctic sea ice and mountain glaciers have been disappearing, precipitation patterns have changed, crop growing seasons and the ranges of many species have shifted.

America's most respected scientific bodies, including the National Academy of Sciences and the US Global Change Research Program, as well as international scientific bodies, state authoritatively that greenhouse gas emissions accumulating in the atmosphere are changing the climate, potentially beyond anything that human civilization has experienced. The scientific basis for this conclusion is firm and no other plausible explanation for the observed changes has been found. Unless greenhouse gas concentrations in the atmosphere are stabilized at levels not far above today's, climate change will lead to weather conditions far different from any experienced in the course of human civilization.

The most threatening consequence of climate change is not the change in average temperatures of a few degrees but the increasing frequency of extreme weather events: heatwaves, droughts, more intense hurricanes, deluges and flooding. These extreme events and their disastrous impacts are becoming more frequent. More of these extremes lie in the future.

Another menacing aspect is the presence of strong positive feedbacks in the climate system. If they are unleashed, climate change could spiral out of control, even if sharp reductions in greenhouse gas emissions are achieved. These positive feedbacks include melting of permafrost, releasing huge volumes of carbon and methane that accelerates the accumulation of greenhouse gases and further warming; release of methane hydrates from the shallow ocean floors of the Continental Shelf, also releasing potentially huge amounts of methane, a greenhouse gas 20 times more potent than carbon dioxide; melting of sea ice, which exposes the northern oceans to the warming rays of the sun; and larger and more frequent forest fires in a warmer world, which releases carbon stored in soils and vegetation. These positive feedbacks are already at work.

Relying in the US on adaptation to climate change would not be an adequate response. The evidence from four or five decades of recorded climate change shows that adaptation has lagged badly and in many key sectors has hardly begun. Adaptation measures have typically been reactive rather than preventive and have been taken only after weather disasters. Adaptation faces severe obstacles, including uncertainty about the future climate, organizational and individual inertia, and weak or perverse economic incentives that often lead to distorted decisions that are likely to result in heavy future economic losses. Adaptation in less developed and more vulnerable countries would be even more difficult, because most such countries lack the capacity to deal even with current problems, let alone future ones. Consequently, national security experts have identified climate change as one of the principal security threats to the US during the 21st century. Reducing these obstacles to adaptation requires that public and private agencies build the probability of climate change into their investment plans, project designs and operating protocols and work vigorously to improve popular understanding of the risks. Otherwise, damages will be more severe.

An adequate response to the climate problem requires a transition to renewable energy and major improvements in energy efficiency that together will lead to an 80 per cent decline in emissions over the next half century. The technologies that now dominate the energy system, including the gas-powered internal combustion engine and the coal-fired thermal power plant, and the primitive electricity transmission and distribution systems are all a hundred years old and must be replaced by better and cleaner technologies. Wind, solar, gas, hydro, biomass, nuclear and geothermal technologies must provide the major portion of electricity generation. Coal-fired power plants without the capacity to capture and sequester carbon must be phased out. Vehicles must become much more fuel-efficient and rely on electric power and liquid fuels derived from biomass. These technologies are already available and technology development is rapid. Deploying these superior technologies on a large scale will not only control the climate problem but will also reduce America's energy dependence, make the energy system less vulnerable to disruption and attack, and improve domestic environmental quality.

The US economy must become much more energy-efficient, eliminating today's enormous wastage of energy. Most of this can be done while saving and earning money. The impediments are neither technological nor economic, but consist mostly of outmoded institutional arrangements and government policies that can and should be reformed. Other prosperous countries use far less energy per dollar of economic output than the US, demonstrating the feasibility of major improvements. Appropriate incentives and policies can also accelerate the emergence of even better technologies in the marketplace.

A far-reaching energy transition over a half century is possible. Such transitions have already happened several times in American history during the past 150 years: from whale oil to kerosene; from animal to motor traction; from water power to steam power; from steam power to electricity. Each time, a new surge of economic growth and innovation resulted.

There are many steps that the government can take to promote and facilitate that transition: policy reforms, institutional changes and re-ordering of investment priorities. Federal energy legislation in 2005,

2007 and 2009, along with numerous State and local government actions, have made a useful beginning and have slowed the growth of greenhouse gas emissions, but that is far from enough. More powerful and comprehensive actions still remain to be adopted if emissions are to be reduced to the extent necessary.

The most fundamental step is to put a price on carbon emissions, thereby allowing the market to respond with innovation and efficiency. A monetary penalty for greenhouse gas emissions would correct a fundamental market failure, in that emitters now do not face any economic cost when their emissions impose additional costs and risks on the rest of the economy. A price on carbon would provide both incentives and maximum flexibility for all firms and households to find ways to improve energy efficiency, conserve energy, replace fossil fuels with clean alternatives, and bring more energy-efficient products and technologies to the market.

The best way to do this is by installing an upstream cap and trade system. Such a system would require permits from all first sellers of fossil fuels – about 2000 companies. Permits would be calibrated to the respective carbon content of coal, petroleum and natural gas. The number of permits made available would decrease over time, reliably ensuring that fossil fuel use and carbon emissions would decline. As fossil fuel availabilities diminished, their prices would rise, stimulating improved energy efficiency and ensuring a market space for renewable. Permits would be bankable and tradable to promote efficiency in fossil fuel use. Although opponents of cap and trade systems now attempt to discredit them, they were introduced in the US by Republican administrations and have proven their value in other domestic environmental policy arenas. They are being adopted for controlling greenhouse gas emissions in European countries, Australia, New Zealand, several Canadian Provinces and many American States.

This system should be extended to reductions in other greenhouse gases and to carbon capture and sequestration by an offset mechanism that awards salable permits to these mitigation actions on a carbon-equivalent basis. Such offsets would greatly lower the cost of the energy transition by taking advantage of low cost opportunities. An upstream cap and trade system could also be linked to international permit markets, including a reformed and expanded Clean Development Mechanism, lowering costs further and reducing permit price volatility.

An upstream cap and trade approach would be intrinsically fair, because higher energy prices would affect all users in proportion to their fossil energy use and their contribution to the climate problem. Burdens on low income households from higher energy prices could be compensated in any number of ways, such as cost of living adjustments to social welfare programs and direct rebates. Other equity claims are far less persuasive. Giving away permits to energy companies and energy-intensive industries would transfer valuable salable assets to corporate balance sheets but would not prevent energy prices from rising. Sheltering coal-dependent States that for decades have had the nation's lowest energy and electricity prices is also questionable. Auctioning permits and using the resulting revenues – $50 to $150 billion per year for many decades to come – to reduce other more distorting federal taxes, to reduce the federal deficit, to finance critical infrastructure investments for the energy transition and to compensate low income households would be a far superior policy approach than giving those permits away to energy companies and energy-intensive industries.

Departures from this approach for political reasons would reduce effectiveness and raise costs. A limited cap and trade system covering only electric power plants would ignore many low cost abatement possibilities and important emission sources. Permit price ceilings would compromise the effectiveness of the system. Limiting the use of offsets would substantially raise costs. If such compromises are adopted at the outset, they would be difficult to correct later on and the higher costs would persist for decades, so it is important to get the policy right on the first try.

All serious economic studies have found that the economic impacts of an upstream cap and trade system and the resulting energy transition would be small and would have a negligible effect on US economic growth. If enacted now, gross domestic product in 2030 would be only 1–2 per cent lower than otherwise and national income would still be at least 50 per cent higher than in 2010. Even these conclusions are too pessimistic because they don't adequately take into account that many energy efficiency improvements would actually save money, that development and commercialization of new technologies would be stimulated, that air pollution, road congestion and other non-climate environmental problems would be reduced, that energy imports would be reduced, that our actions would stimulate corresponding actions in other emitting countries, and that if this energy transition does not take place, climate change damages and risks would be increasingly severe.

The energy transition would create jobs, on balance, because the rapidly growing alternative energy industries are more labor-intensive and have a smaller import content than the fossil energy industries that would shrink. International trade impacts would be small and confined to a few industries, because America's main trading partners, including China, are also taking actions to reduce greenhouse gas emissions. However, if the US does not soon take action, those countries will capture a possibly irreversible advantage in wind, solar and other clean-energy sectors.

International cooperation is essential, because neither the developed nor the developing countries can stabilize the world's climate by acting alone. American policy is critical, because other nations will not act unless the US, the world's richest and second largest emitter of greenhouse gases, does so. Negotiations with the developing countries must acknowledge that economic growth is essential to their efforts to reduce the burdens of poverty, just as those countries have recognized their 'common but differentiated responsibility' to help reduce emissions. China has already carried out significant policies and measures to meet this responsibility.

The next step in international negotiations would balance commitments in developed Annex I countries to achieve ambitious targets and timetables for emission reductions against commitments in non-Annex I countries to specific policies and measures that will reduce emissions while simultaneously contributing to economic growth. There are many such win–win possibilities in developing countries, including major improvements in energy efficiency, reform of energy subsidies and reduced deforestation. Another area of potential agreement is reform and expansion of the Clean Development Mechanism by making the baseline for crediting emission reductions less uncertain.

Despite the urgent need to start reducing emissions in the US, adequate action by Congress is doubtful in the coming few years, because of the political influence of fossil energy industries. Action in the Senate is particularly difficult to achieve, because of possible filibusters and the disproportionate weighting of votes toward resource-dependent States. Public understanding of the climate problem and demand for action is still not sufficient to force reluctant legislators to act against the interests of politically powerful interests.

Nonetheless, the administration can help build public support for action by emphasizing the link between climate change and the increasing frequency of extreme weather events and weather disasters. It can also put pressure on the electric utilities, the coal industry and their Congressional allies to agree to an equitable policy package by using its current regulatory powers to promulgate climate change regulations and to tighten controls over coal's many other environmental impacts.

If climate change is to be kept within tolerable limits, US emissions must start declining in this decade, so there is no time to spare. Because positive feedbacks may be unleashed that will make it difficult or impossible to control the extent of future climate change, actions by current leaders are critical. This book has indicated a way forward that can resolve the most serious environmental issue of the 21st century while simultaneously ensuring future prosperity. If this becomes the road not taken, it will truly make all the difference in the world.

Index